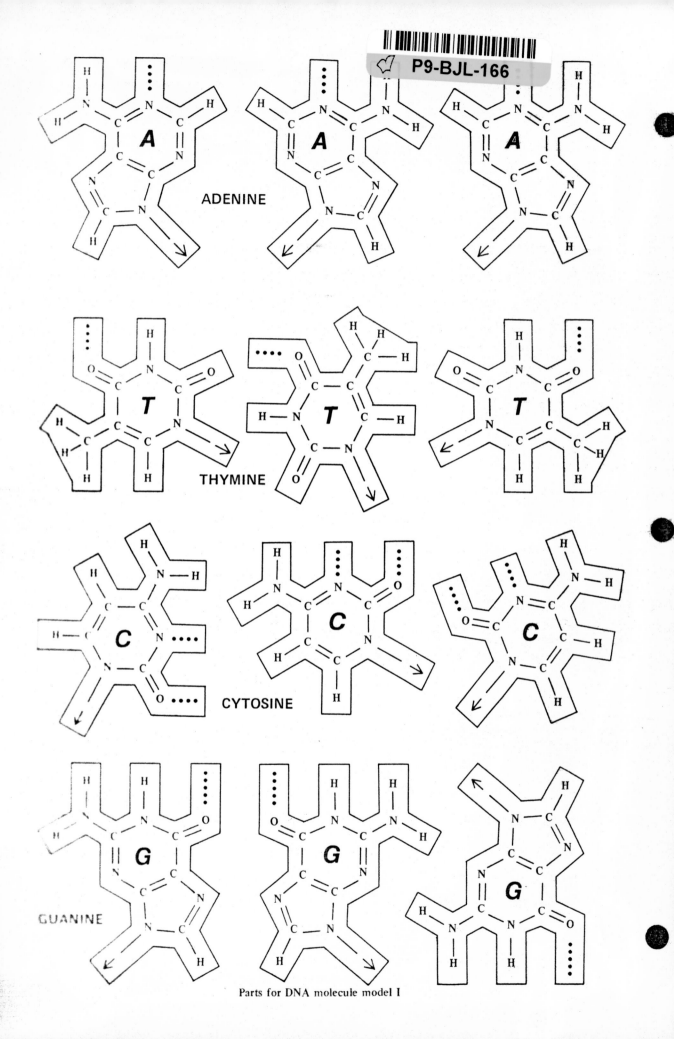

ADENINE

THYMINE

CYTOSINE

GUANINE

Parts for DNA molecule model I

PROBLEM SOLVING IN
BIOLOGY

PROBLEM SOLVING IN
BIOLOGY

THIRD EDITION

INCORPORATING EXPERIENCES IN LIFE SCIENCE

A Laboratory Workbook

Eugene H. Kaplan

Hofstra University

Macmillan Publishing Co., Inc.
New York

Collier Macmillan Publishers
London

Macmillan Publishing Co., Inc.
866 Third Avenue, New York, New York 10022

Collier Macmillan Canada, Inc.

ISBN: 0-02-362050-1

Printing: 1 2 3 4 5 6 7 8   Year: 3 4 5 6 7 8 9 0

ISBN 0-02-362050-1

Preface

Eleven years have elapsed since the publication of the original *Problem Solving in Biology* and *Experiences in Life Science*. I have learned a lot in the intervening years from the comments of colleagues and the responses to questionnaires sent to faculty using the books all over the country. Most wanted crisp, straightforward exercises that were more easily graspable by the students. Many of my more ambitious efforts such as an exercise on exobiology, and the series of replications of famous experiments, were not used. Since the requests for straightforwardness came from the users of both books, it seemed logical to combine them into one. There are more than enough exercises included in this manual for either the one semester or the two semester course. Only the most practical exercises of both books have been retained, and several new chapters have been added — one where the student develops Koch's postulates for himself. A "universal" field trip has been included as an exercise in ecology. It deals with the rotting-log community. It is "universal" in the sense that this is the only biotic community available in urban areas, where any vacant lot can be seeded with wooden boards or logs — and it provides an interesting, self-contained, easily studied community even in rural areas where other, more obvious ecosystems require more complex, more massive efforts.

Virtually every exercise has been revised and modernized. The five kingdom system of classification has been adopted; cellular ultrastructure is covered.

But the focus of this laboratory manual remains unique — the requirement that each exercise must present the student with a biological problem to solve by *himself or herself*. Each exercise has two titles, two subjects, two goals. One is the learning of a certain amount of content — this goal is common to all biology books. The other is to come to grips with the epistemology of science. For biology is not the sum of its facts — it is a dynamic process whereby those facts have been wrested from nature by persons who specialize in applying a special attitude and intellectual process toward the biological world. To present biology as a set of facts to be learned by rote — facts that are stated by an authority figure — the all-knowing professor — thundered from the pulpit as it were, is to cheat the student. For the drama of biology often lies beyond the challenge of understanding complex ideas — it lies in the feeling of discovery when the biologist finds out that his hypothesis is correct.

Each of the exercises in this book gives the student the chance to feel the thrill of successfully solving a problem. To the student each solution is a real discovery — for he is repeating the age-old scientific quest — he is finding out for himself.

Many people have contributed to the development of this book — too many to acknowledge individually. My special thanks go to the dozens of students who struggled, good naturedly, together with me, to help make the text more comprehensible and interesting.

A number of colleagues all over the country took time to fill out questionnaires I sent to them to help set the direction of this book. Their experience with previous editions of *Problem Solving in Biology* and *Experiences in Life Science* made their comments invaluable. I am grateful for their assistance. In addition, the following colleagues reviewed the manuscript and made helpful comments: Garland Allen, Washington University; Robert Barkman, Springfield College; Peter Bellinger, California State University, Northridge; Michael Grant, University of Colorado; Jerry Wermuth, Purdue University.

Gary Grimes made the photographs of human tissue types and contributed the plates depicting the ultrastructure of the cell, which were made in his laboratory by Carleton Phillips. Seymour Leicher developed and printed the fetal pig and circulatory system plates. Jeannette Schneiweiss and Ruth Belkin made useful comments and corrected errors in *Problem Solving*.

My wife, Breena, and daughters, Julie and Susan, shored me up against the vicissitudes encountered in developing this book. In addition, Julie and Jacqueline Damsky typed the manuscript. Susan contributed the illustrations for Chapter 33. My heartfelt thanks go to the above-mentioned people.

This book is dedicated to the memory of the master biology teacher of our generation, Jacob Bronowski.

E. H. K.

Contents

PROBLEM SOLVING IN
BIOLOGY

To the Student

When you graduate you will enter a world where course grades will no longer be an indicator of success or failure. Instead, the same forces of selection which affect the survival of every animal species will operate. If you are "fit" (a Darwinian term referring to one's ability to function adequately in one's environment), you will succeed. But fitness no longer is synonomous with brawn (in fact, it rarely was). In our society, fitness is measured by versatility, flexibility, cleverness, creativity − in short, the ability to think and work effectively − to cope with problems. This laboratory program teaches more than traditional, descriptive biology. It is aimed at helping you to *function*. Each week you will be presented with a problem to solve. Each week you will have a chance to practice the kinds of behaviors you will be employing in the "real world." You will be able to ask for the opinions of your classmates − just as you will have the help of your colleagues on the job − *but there will be no "Big Daddy" in front of the room to tell you the answers.* You are fully responsible for solving the problems yourself. Your instructor will give much encouragement and minimal help − an occasional word if you are way off the track, a little help with a tough dissection − that's all.

You should welcome this approach since it is closer to the reality you will face after your four-year undergraduate training period. This course alone will not teach you how to think, but together with your other undergraduate experiences, it will prepare you for the future, where ability to function − to plunge into a problem and get your feet wet − to work with your hands as well as your mind − to face life's problems undaunted − will spell success.

I hope you will welcome your laboratory experience. It will be a relief from the unremitting, detached cerebration of your lecture courses. Here, at last, you can come to grips with problems. You can smell them and feel them − and experience the triumph of seeing the solution unfold before your eyes.

I Introduction

1 What Is Science?

Science is the systematic accumulation of knowledge through the use of logic based on factual evidence. Science is at one and the same time an organized body of facts and a method of problem solving. This relationship between information and the means of its accumulation will occupy us for a considerable portion of our laboratory experience in this course.

A problem is presented below. Your instructor has supplied you with a blank sheet of paper. Try to solve the problem and write down the thought processes you employed (the ideas you had) in determining your solution.

■ THE PROBLEM: *You have been given a closed cardboard box. Your task is to determine what is in the box* without opening it.

Do whatever you feel is necessary in your quest for the answer—short of opening or damaging the box. (Are you at any greater disadvantage than a nuclear physicist who is unable to see into the "box" [atom] in which are the particles that must be liberated in order to cause an atomic reaction?) Solve the problem, remembering to write down what you are doing and thinking **before you read any further.**

Science is a method of bridging the gap between what exists in reality and its image in the mind of man. It strives to relate the intelligence of man to his environment so that his interpretation of his surroundings will be accurate and valid. The assurance that his perceptions are relatively precise representations of what he observes enables the scientist to conjecture about what he has perceived, *with his observations as the foundation for his reasoning.*

The basis of all scientific thought, then, is accurate observation. When an observation is made in accordance with the restrictions inherent in science—that is, when it closely approximates that which it purports to describe—it is considered a fact.

Let us analyze the concept of "fact" and the limitations which it imposes, for this concept is at the very foundation of science. Suppose that you have a branch from a willow tree; let it drop from your hand and you might notice that it falls on a moist spot on the ground. You pick up the stick, and again it falls, once more landing on a wet area. If, on the basis of your observations, you remark that the stick seems to be attracted to water, you will have described what you have *seen.* You will have used your sense of sight to establish a relationship between an environmental phenomenon and your mind. But shall we consider your observations as fact and include them in that network of "valid observations" from which all scientists draw the substance of their reasoning? In other words, are you willing to accept the report of an individual's perceptions as absolutely valid without any further investigation? To do so would be to imply that man is incapable of making an incorrect observation, that one's senses never lie. But you have no doubt had the experience many times, perhaps not without embarrassment, that your observations have been proven incorrect.

It has been shown that error can lie either with the sense organs (two people look at a light, one reports that it is red, the other, who is color-blind, that it is gray), or with the more complex aspects of the nervous system, the integration centers ("thinking" area) in the brain. We are all familiar with instances where the mind projects information into the content of what is sensed. For example, a report of a fight between two men might be described completely differently by persons **prejudiced** in favor of one or the other. The very word "prejudice" (meaning prejudgment) demonstrates our acceptance of the concept that man is prone to modify what he senses along certain preconceived lines. It is important to realize that there are numerous causes for error in observation—*many of which are not apparent to the observer*.

Let us return to your report on the willow branch which seems to be attracted to water. We have suggested the objections to the acceptance of this report as it stands. To what conditions must we subject the report in order that we may be able to accept it as true—that is, as an unbiased representation of what has occurred?

1. What must be done by the observer before his observation can qualify as fact? _____

_____

2. What must be done by other persons interested in the accuracy of the observation

before they can accept it as fact? _____

Your answers to the preceding questions should have indicated ways and means of making what is observed **empirically verifiable**—in other words, provable by the senses of others. In order for us to accept an observation as fact, we must provide for corroboration by others. If twenty people hold up the willow branch and release it, we can accept their affirmation of your observation under the assumption that not all of them are subject to the same bias which influenced you. Situations which occur only once, or situations not observed by several people, cannot be considered as facts.* The same is true of statements of belief which do not permit corroboration by the senses.

If you are not entirely satisfied with your answers to questions 1 and 2, reread the foregoing paragraph and try again below.

1. _____

_____

2. _____

_____

3. Now list areas which are not susceptible to scientific reasoning and explain why they are not:

a. _____   _____

_____

_____

*It is possible to consider events observed only once as facts, but we are less willing to believe in their existence without further proof. We say that such facts have a low level of validity.

b. _____   _____

_____

_____

c. _____   _____

_____

_____

d. _____   _____

_____

_____

In summary, science is concerned only with aspects of the environment which can be sensed and which are empirically verifiable.

The accumulation of related acceptable facts is the end result of the scientific process and not the *modus operandi* (way in which it works) of science. In other words, facts are only the bricks with which the scientific edifice is built.

4. If the goal of science is not merely to accumulate facts, what is its essential function?

(Suggestion: Reread the first paragraph of this chapter.) _____

_____

_____

HOW SCIENCE FUNCTIONS

Does science attempt to determine what is "good" and what is "evil"? Does science have a moral purpose? Does science attempt to convince anyone of anything? In evaluating these questions, remember that the scientist must confine himself to the building of new knowledge on the framework of "older" confirmed observations. His task consists of:

(1) Understanding what has already been accepted as fact.
(2) Suggesting relationships among these facts or (from these relationships) suggesting the existence of *new* facts.
(3) Testing the validity of these suggested relationships.
(4) Presenting the new facts in a manner which reasonably insures that there is no bias on the part of the investigator.

5. What is the best way for others to determine whether or not the relationships

suggested by the scientist are valid?_____

_____

THE SCIENTIFIC METHOD

The logical processes just outlined comprise those thinking procedures which are generally grouped under the term "scientific method." Let us follow the reasoning employed by a person wishing to solve a problem susceptible to scientific analysis. First, however, it should be noted that

(1) Such a person may be anyone—scientific thinking is not restricted to white-haired, white-coated thinkers in "ivory towers."

(2) The problem may be *any* problem which begins with fact—not only problems restricted to work with elaborate instruments such as microscopes or cyclotrons.

(3) The aspects of scientific thinking are placed in sequence as though several processes were occurring in strict succession. This is simply for convenience in explanation. In practice, there is no one set of steps which, when followed, will contribute to the solution of all problems. There are, however, certain *general* processes which are involved in logical thought. These will be explained below.

■ THE PROBLEM: *A person wishes to buy a new car. He would like to find out which car is the "best" one.*

From previous experience with new automobiles, he is partly committed to "brand X." He mentions his belief that a brand X car is "best" to a friend who laughs at him, saying, "Obviously, brand X is not the best . . . . It has half the power of brand Y and is much less roomy than brand Z."

It is apparent that the two individuals involved in this hypothetical situation cannot communicate meaningfully with one another because each has his own concept of what is meant by "best" in relation to automobiles. Thus the first obstacle to valid reasoning has presented itself. Leaving our two disputants locked in an argument which may last for hours, we will offer an **operational definition** of "best" auto. For our purposes, "best" will mean greatest fuel economy and ease of handling. The importance of this definition will become apparent later on.

Thoroughly confused by the argument with his friend, our prospective purchaser decides to ignore all of his previous considerations as to what is desirable in an automobile and to examine each brand of car without making any decision as to its respective merits, hoping that when he finishes his investigations the proper choice will be apparent to him. After listening to ten different salesmen extolling the virtues of their cars, he goes home expecting the answer suddenly to occur to him, but it does not. Each car has certain "good points" which were stressed by the salesmen; so all our buyer is left with is a collection of arbitrary virtues, and a headache. He has made no attempt to tie his information together, no attempt to pinpoint any of the data. The result is further confusion—especially when one salesman pointed with pride to the large size of his car and another extolled the virtue of the "compactness" of his car.

The prospective purchaser has made a fundamental error in his reasoning. He violated the **law of parsimony.** The aim of this law is to help to conserve intellectual effort. It says, in effect, that the simplest means for solving a problem lies in excluding all extraneous facts, explanations, causes, etc., and that the simplest answer which will account satisfactorily for all the known facts is the preferred one. In this case our automobile buyer did not focus his investigation on a specific objective. He stared blankly at each car expecting something to happen, just as some biology students stare into the microscope without any idea of looking for some specific thing, hoping that the solution to their problem will appear, as if by magic, out of the welter of shapes and sizes before them.

By trying to consider everything, our friend resolved nothing; he erred in obtaining too many unspecialized data. The reason for his overgeneralized approach was that he had no guide to help him to pinpoint his thinking. It was important at first to obtain a certain amount of information by random observation, but such material could not lead him to any constructive thinking until a *particular relationship* was clearly defined between some of the aspects of this random information. (This becomes more obvious when you look for a friend in a crowd; either you stare futilely at everybody or you look for the hat or suit your friend is wearing.) Recognition of the significant information makes it possible to suggest a relationship between one aspect of the data and another.

In this case our friend might have suggested that either brand A, B, or X would have been the car to buy because they were all small, economical, and easily maneuvered automobiles. Such suggestions or ideas, aimed toward the solution of a problem, and based on previously obtained facts, are called **hypotheses**. A hypothesis is a suggested *possible* solution to a problem based on fact. It is differentiated from a guess by its reliance on observation (fact) for its origin.

6. Since a hypothesis is only a suggested solution to a problem, what must be done to it

in order to arrive at a valid solution? _____

_____

All suggested hypotheses must be tested in order to determine which one is the most acceptable solution of the problem. The test of a hypothesis is usually called an *experiment*. In order to test the hypothesis that brand X car is the best, the buyer would have to drive the car. Is driving brand X car sufficient to determine whether it is the best—that is, better than cars A and B?

7. Improve upon the experiment so that it will adequately determine which of the

hypotheses is most valid. _____

_____

_____

_____

_____

By driving all the cars under exactly duplicated conditions, one would be able to determine which is best. To obtain the most valid results, the performance of each car should be compared to that of a constant standard. Thus, if the subject drove his own car on a measured course when the wind was behind him at 12 mph and the temperature was 74°F, and then drove cars A, B, and X under exactly the same conditions, he would have a valid basis for comparison because he would have eliminated other **variables** or influences which might have affected the performance of each car. He would thus have compared each to a known standard.

The number of variables must be regulated or limited, so that ideally only one possible cause may be attributed to an effect. The means by which the variables are controlled is called the **control**. To illustrate: If you wish to test the hypothesis that light affects plant growth, you take two groups of plants that are more or less identical. One group is given light and the other, the control, is kept in the dark; all other conditions—food, soil, etc.—are made the same (equated) for both groups. If the plants in the group given light grow more rapidly, then you are sure that these plants did not grow faster because they had more food, better soil, etc., since they were subjected to exactly the same conditions as the control group which did not grow rapidly. The only possible cause (of those apparent to the investigator) is the light, because all other variables (causes) were controlled. The hypothesis has now been adequately tested and validated.

The story about the search for the "best" car has a happy ending. Upon completion of this experiment, our buyer concluded that car X was the "best" and purchased it—and of course this was the right choice (all other things being equal) as we can see by his data.

	Miles per Gallon	Ease of Handling (Turning Radius) (feet)
Car A	22.3	31
Car B	23.6	30
Car X	25.4	29
Present car (control)	18.4	34

This decision was based on a hypothesis which had been proved valid. Such a decision is called a **conclusion**.

If, however, our buyer had started with the hypothesis that car B was his best buy, he would have found from his data that his hypothesis was **invalid**. His conclusion, therefore, would have been based on a hypothesis proved invalid. It can be seen, then, that a conclusion is based on hypotheses which would have been proved valid or invalid.

RE-EXAMINATION OF THE PROBLEM

Turn your attention once again to the closed box. Try again to determine what is in the box, this time using the elements of the scientific method. State specifically all **hypotheses**, **experiments**, etc. Use the back of the sheet of paper which contains your original effort. Write a paragraph of not less than fifty words comparing your two efforts. Let your answer take the following questions into consideration.

a. Do we use elements of science in our everyday thinking?
b. Did the scientific method help to uncover the answer to the problem? How?
c. Must we always use scientific thinking when faced with a problem? If so, why? If not, when should we use this method?

ANOTHER VIEW OF SCIENCE IN MODERN SOCIETY

So far you have learned that the scientific attitude has brought about a transformation in our understanding of the natural world. Science has allowed humankind to do what no other organism has been able to do—to control its environment. No longer are we as subject to the ravages of storms, hunger, thirst, disease, and other natural forces as we once were. In the past they threatened us with extinction. Science and humankind have flourished together. Today we are at the pinnacle of this relationship. Nuclear power plants may provide us with unlimited energy—genetic engineering may provide us with substances which will control disease to a once-unimagined degree.

But science has become a huge monolithic institution in our society. It has become so sophisticated and has so much potential power that it is almost impossible for the lay person to fathom the extent of its activities. Scientists themselves must often view with awe the forces which they can unleash. Since they are human, scientists can make mistakes. Witness the remorse and sudden social awareness exhibited by some of the scientists who developed the atomic bomb in the 1940s.

Lately there has come to exist an uncomfortable feeling among many of us that science is not a source only of universal good. It can be harmful on a scale which dwarfs past dangers to our species. The following papers were taken from a Nobel Institute symposium examining the role of science in modern society.

People, Knowledge, and Science—*Jacqueline Feldman**

Jacqueline Feldman received her doctorate in Theoretical Physics. Later she became interested in the study of mathematical methods in the social sciences, and is currently a researcher at the Centre National de la Recherche Scientifique at the University of Paris-Sorbonne. The portion of the article reprinted here comes from a paper dealing with the different positions of men and women in the institution of science, from a feminist point of view.

Let us first take a look at science, as it began in the seventeenth century, when a few fervent people began to interpret the world in a new way. Of course, a certain kind of knowledge existed already; however, it is difficult for us today to imagine what this knowledge meant to people, because we have learned from childhood that scientific knowledge is the only valid one. It represents the real truth and fights against obscurantism.

This current ideology, this imperialistic attitude, I shall call *scientism*. It is, for example, difficult for us to imagine what knowledge meant to Newton, who was also deeply interested in alchemy and theology; scientism appears to have taken from Newton that which interested it and left out that which made up his humanity.

A lot has been written about the scientific method. My own experience tells me that it is less a question of *method* than of *attitude*. The current emphasis on method rather than on attitude is meaningful. A method can be put into rules and communicated to many people, who learn that this is what is called science. The rule then becomes a dogma.

Attitude, on the other hand, first concerns people; and transmission of the scientific attitude occurs by direct relationships between people.

This was indeed the case when science developed, but it cannot exist in the huge institution we know today.

Scientism makes us forget that the first reality of science is people, not rules.

This perversion of the scientific attitude by the scientific institution was pointed out before World War II, by Bernal, a Marxist scientist. I can do no better than reproduce his words, still valid today.

As to the learning of scientific method, the whole thing is palpably a farce. Actually, for the convenience of teachers and the requirements of the examination system, it is necessary that the pupils not only do not learn scientific method but learn precisely the reverse, that is, to believe on the authority of their masters or textbooks exactly what they are told and to reproduce it when asked, whether it seems nonsense to them or not.

*From *Ethics for Science Policy—Report from a Nobel Symposium*, Pergamon Press, Elmsford, N.Y., 1979. © 1979 The Nobel Foundation.

At that time Bernal still believed that science, if well developed—for example, in a socialist country—could be equivalent to progress for humanity.

Today, this perversion can no longer be looked upon as a mere accident and must be thought of as having its roots in science itself and in its intimate relation with the industrial societies.

We all know that the technical achievements of science have themselves led us to the atom bomb and to biological weapons; they have also led to pollution, and to the destruction of the natural environment. A certain rationality inherent in science was an integral part of Nazism. Other kinds of rationality are producing huge towns where more and more young people are driven to delinquency. Social and psychological sciences now tend to expropriate the knowledge that people have about themselves. Psychiatry is used as a means of doing away with political dissenters.

All of this concerns science, and our first task should be to understand better how the great and generous dream of the last century has failed us and led us where we are, into a world of confusion, of violence, of distress.

The Fragmentation Principle

A first answer is given by the observation that some kind of fragmentation principle is inherent in science.

Science has made its greatest achievements by dividing reality into pieces and concentrating all of its efforts on one of those parts. This has sometimes produced—as we all know—a great deal of power. But the same fragmentation principle has led the scientist to a total irresponsibility with respect to the use that is made of that power. He is an expert on only a tiny piece of reality, and others decide what to do with the power.

The same fragmentation principle has divided society into the scientific community and the others. As the knowledge of this community increases, the division becomes greater, and people on the outside become more and more helpless and ignorant, more and more dependent upon experts.

The fragmentation principle applies not only to research and to society, but it also applies to the scientist himself. I have already mentioned the total irresponsibility incurred in him: all he has to do is to obey the rules of the scientific institution. His highly praised creativity is channeled along those lines. His intelligence is separated from his emotions.

Science, Knowledge, Wisdom

We have become so used to this kind of piecemeal knowledge that we have great difficulty in imagining that it could be different. I would just like to mention that in other civilizations, knowledge is linked much more strongly to the other aspects of the personality and helps to unify the person. Eastern disciplines, such as yoga and za-zen, bring to the people who practice them a deeper knowledge of themselves and tend to unify rather than divide the self.

I was thinking recently of the way in which science, knowledge, and wisdom come together. Science could be seen as being only one means of obtaining knowledge, which should help us to acquire wisdom: we human beings have a deep need for some kind of poor man's philosophy to help us to live in this difficult world. This seems to have been the case in many of those societies which are outdated today and sometimes described as "archaic."

My office is in a modern building called the Maison des Sciences de l'Homme, literally translated, "the House of the Sciences of Mankind." There is a fine library. I looked up the word "science" in the subject file. There were about a thousand index cards under this word: all of the books that have been written on the scientific method, the different sciences—exact, human, applied—and their histories. I next looked up "knowledge": there were only a few dozen index cards, because there is a very official sociology of knowledge. And then I looked up "wisdom": there was nothing, not one single card.

So, after two thousand years of Christian civilization, four centuries of rationality, two centuries of industrial society, in the century of the boom of the scientific institution, wisdom is a completely outdated notion.

Science, which was supposed to bring wisdom to people, has totally forgotten its original aim and has become a kind of crazy logic machine, developing for its own sake and escaping humanity.

An Imperialistic Institution of Knowledge

Scientism recognizes only one type of knowledge: that based on science.

The scientific community acts as a warrant for science—that is, all of those who are recognized as scientists by scientific institutions. But what was possible when science was animated by a few fervent people—fighting for new kinds of knowledge—can no longer exist now that science has become more than a respectable institution, the single source of recognized knowledge.

The only guarantee of scientific work of good quality is the deep involvement and devotion of the scientist to his work. The handing down of the scientific attitude is possible because of an existing human relationship: that of master and student. This relationship is not given a chance to exist and develop today.

Nowadays, a scientist chooses a career. He does not follow a vocation. The institution has developed its own criteria of social success. Showmanship has taken the place of reality. The number of publications is more important than their content. Participation in international conferences or symposia takes over the role of public relations. Intelligence consists in choosing the right research director, the right laboratory, and the right country, and in sensing which subject is going to become fashionable.

A director no longer has time to work with his many students. The number of students he has serves, firstly, his own prestige, and no one cares about the relationship of those students to knowledge or the human relationship between the director and his students.

In a sense, science no longer exists. There is only an imperialistic institution of knowledge, based on the scientist ideology.

Scientism's Strategy

Scientism uses the real achievements of some parts of science to defend its extravagant claim that knowledge can only be begotten from science. I would like to show how the scientist ideology is spread, by material from a book called *The Scientist*, one of a collection on science edited by Time-Life. At one point, the esotericism of science is discussed. The authors report criticisms of various pedantic expressions, used where simple language would do. The criticisms refer to the medical and social sciences, and the authors write, "the scientific words and sentences are often cumbersome but they often have reason on their side." However, the authors then take their examples from the fields of physics and chemistry.

The strategy is clear. In order to justify the absolutely unnecessary jargon used in the medical and social sciences, the authors refer to two fields which utilize not jargon but necessary and precise language.

The same strategy runs through the whole book: that use of the scientific method is unavoidable is illustrated by physics, for the good reason that this is the only field in which it really applies.

In the same order of ideas, mathematics plays a very particular role in science today. It represents the esotericism of science itself, it is the very language of our dehumanized civilization. The uses and abuses to which it is put again pervert this beautiful creation of the human mind.

The more a discipline lacks pure scientific status, the more it will tend to justify itself with an appearance of scientism; and mathematics is the perfect expedient.

Social sciences use mathematics as best they can—which is rarely the best they could—to develop a kind of knowledge that has little to do with real science. The label "science" is used to give a legitimacy to yet more experts on the subject of mankind.

Science Against Knowledge

Science, as any other institution that has become dominant, finally hinders its own aim. The analyses of Illich of the present state of industrial societies are perfectly applicable here. Illich takes the example of traffic: cars have been very ingeniously invented to enable us to move faster from one place to another. But the traffic situation is such that one often sees for oneself that one would go faster with a bicycle. Another example extensively studied by Illich is that of the medical world. Here, the official aim is people's good health. But by putting all the emphasis on experts, and a few spectacular successes, our relationship to health has deteriorated profoundly. We have been changed into passive patients who have lost much of our natural combativeness against illness.

The notions introduced by Illich fit the case of science perfectly: the *radical monopoly of science* makes us more and more dependent upon experts, whereas the institution itself produces less and less work of quality. In the same way that medicine produces new diseases (iatrogenic diseases), science is now producing new kinds of ignorance.

Knowledge for the People

Scientism does not, in fact, like people; this is obvious in the social sciences, where, in order to know about people, that is, about ourselves, they are made into objects of study, numbered, and filed.

The lines along which I would like to think about knowledge are lines which, in our own, western way, will reconcile people with themselves, science with emotion. This is not, I believe, Utopia. It is the next step to be taken after science. I am not talking about a return to the past; I am saying that rationality and science have led us far enough into a fragmentation of our personalities and of our society. Now, we can strive toward a reunification of our personalities. Fragmentation might have meant progress in the seventeenth century. Now it means danger and impoverishment.

Nowadays, the idea of "quality of life" is replacing the idea of development. Rather than developing the scientific institution beyond all bounds, we should be preoccupied with the quality of knowledge, for the individual and for society.

Science as a Source of Political Conflict—*Dorothy Nelkin**

Dorothy Nelkin is a professor in the Program on Science, Technology and Society and the Department of Sociology at Cornell University. She has published a number of books on politically controversial areas of science and technology, and on citizen participation in decisions concerning technology.

"Science Finds—Industry Applies—Man Conforms"; that was the theme of the Chicago World's Fair of 1933, celebrating a century of scientific and technological progress. The relationships expressed by this theme remained largely unquestioned during the period of rapid economic growth that followed World War II.

But belief in technological progress has since been tempered by awareness of its ironies. Technological "improvements" may cause disastrous environmental problems: drugs to stimulate the growth of beef cattle may cause cancer; "efficient" industrial processes may threaten the health of workers; beneficial medical research may be done at the expense of human subjects; and a new airport may turn a neighborhood into a sonic garbage dump. Even efforts to control science and technology may impose inequities, as new standards and regulations pit quality of life against economic growth and the expectation of progress and prosperity. Thus, the relationship between science and the public is now marked less by conformity than by persistent political conflict and a search for controls.

The past decade has been remarkable for the amount of political action directed against science and technology. Issue-oriented organizations form to obstruct specific projects, and many groups demand greater accountability and public participation in technical policy decisions. Even scientific research has lost its exemption from political scrutiny: an issue of *Daedalus* on "the limits of scientific inquiry" examined the proposition that some kinds of research should, in fact, not be done at all. A conference on the "social assessment of science" examined international efforts to impose regulations on research. A political scientist wrote about the "crisis" in science, "attacked from all sides [by] a significant coalition of reactionary and left-wing thinking."

Science has always faced ambivalent public attitudes. The acceptance of the authority of scientific judgment has coexisted with mistrust and fear, as innovations such as vaccination and research methods such as vivisection have revealed. The romantic view of the scientist as "a modern magician, a miracle man who can do incredible things" parallels the negative images of

Dr. Faustus, Dr. Frankenstein, Dr. Moreau, Dr. Jekyll, Dr. Cyclops, Dr. Caligari, Dr. Strangelove. . . . In these images of our popular culture resides a legitimate public fear of the scientist's stripped-down, depersonalized conception of knowledge—a fear that our scientists will go on being titans who create monsters.

Even as, in recent years, attacks have mounted, surveys of public attitudes suggest that science is seen favorably as instrumental in achieving important social goals. About 70% of Americans believe that science has changed life for the better. Furthermore, the standing of scientists relative to other occupations has improved continually: in the United States, scientists rank second only to physicians in occupational prestige (they were fifth in 1966). Similarly, in Europe, a survey by the Commission of the European

*From *Ethics for Science Policy—Report from a Nobel Symposium*, Pergamon Press, Elmsford, N.Y., 1979. © 1979 The Nobel Foundation.

Communities found widespread consensus that "science is one of the most important factors in the improvement of daily life."

What, then, is the significance of the flare-up of disputes over science and technology? Are recent controversies a manifestation of the "crisis of authority" associated with the 1960s, or are they simply local protests against decisions that affect particular and immediate interests? Do these disputes express widely shared ideological and political concerns, or do they simply reflect traditional anti-science sentiments and resistance to technological change?

The way in which one perceives science and technology reflects nontechnical considerations—special interests, personal values, attitudes toward risk, and general feelings about science and technology. The threat to human values may assume far greater importance than any details of scientific verification. Perceptions therefore differ dramatically:

Is recombinant DNA research a potential boon to medical progress or a risky procedure continued only because of vested interests among scientists?

Is nuclear power a solution to the energy problem or a destructive force perpetuated because of existing industrial commitments?

From one viewpoint, the activities of protest groups resemble nineteenth century Luddism—a wholesale rejection of technological change. Zbigniew Brzezinski calls such opposition "the death rattle of the historically obsolete." But from another perspective, protest is a positive and necessary force in a society that, Theodore Roszak claims, "has surrendered responsibility for making morally demanding decisions, for generating ideals, for controlling public authority, for safeguarding the society against its despoilers." Thus, while most of us are "frozen in a position of befuddled docility," protest groups fight to preserve the values lost in the course of technical progress.

Concerns About Science and Technology

During the spring of 1977, at a forum on recombinant DNA research, protest groups invaded the austere quarters of the National Academy of Sciences singing, "We shall not be cloned." During the same spring, thousands of protesters gathered at a reactor site in Seabrook, New Hampshire, to block the construction of a nuclear power plant. Biomedical scientists doing experiments on the effect of antibiotics on the human fetus have been indicted for "grave robbing." Public pressure forced the termination of a Harvard Medical School Project to test the association

between certain chromosomal abnormalities and a predisposition to antisocial behavior. And controversy ended a research project on the effect of psychotropic drugs on behavioral disorders. Citizens have objected to science-based programs such as genetic screening or swine-flu vaccination, and some have even objected to the teaching of evolution in public schools.

Many of these disputes developed out of fear of potential health and environmental risks, such as those associated with nuclear power or recombinant DNA research. But some reflect growing uneasiness about the social uses of knowledge: the fear that research findings may be used for harmful ends. Other disputes occur because people consider scientific research morally dubious, a threat to their cherished beliefs. Questions of equity or justice arise over the allocation of resources or the distribution of economic and social costs. Finally, science and technology often appear to infringe on individual freedom of choice.

The Fear of Risk

The technique of recombining DNA molecules has opened up one of the most promising areas of contemporary scientific research, with potentially enormous benefits to society. Yet this new technique, it is feared, might also inadvertently create genetic changes in known pathogens, or produce dangerous new forms of microorganisms for which humans have no resistance and medical science no cure. This possibility, however remote, could be catastrophic and has been the source of prolonged and heated public debate.

We are deluged with warnings about cyclamates, polychlorinated biphenyls (PCBs), freon, toxic substances, and radiation—the list is long and growing, so the fear of risk is inevitable. Moreover, DNA research is not the only area of scientific investigation that poses problems of risk: similar questions are raised about cancer research using human tissue and about chemical research with poisonous substances.

But assessing risk is complicated: often, although an accident could be catastrophic, the chances of one happening are small and difficult to calculate. In the case of nuclear waste disposal, it is the fear of an unlikely but potentially devastating catastrophe that sustains conflict. In other cases, risks are known but must be weighed against potential benefits; in this case dispute focuses on balancing competing priorities in decisions about regulation (for example, in setting worker safety standards).

The recombinant DNA dispute shares many characteristics with the nuclear debate. In both, it is the possibility of unlikely but devastating catastrophe that provokes concern. In both cases, public fear is exacerbated by the "invisible" and only vaguely understood nature of the risk. How does one know if a lethal gene has been produced, or if a nuclear waste storage facility is adequately protected against radiation leakage? In both cases, uncertainty about risk leaves open large areas for conflicting interpretation. Disagreement among scientists increases public confusion and doubt by revealing a limited ability to predict the impacts of science and technology. Finally, in both cases, value-laden questions intrude: what is an acceptable level of risk? Who should assume responsibility for evaluating science projects when they may have an impact on public health?

Increased evidence for the safety of research often fails to allay public concern, because factors other than immediate risk have served to generate conflict over science and technology.

The Fear of Misuse

In 1975, an acrimonious dispute took place over a research proposal by Harvard University scientists to study male children with an XYY chromosome pattern. The proposal grew out of a hypothesis about the relationship between this genetic aberration and criminal behavior. This hypothesis was based on observations of a high incidence of the extra Y chromosome among men in prisons. The investigators proceeded with a chromosome-screening program, hoping that better understanding might facilitate the development of remedial and therapeutic intervention. Critics claimed, however, that XYY research was more likely to perpetuate damaging assumptions about the genetic origins of antisocial behavior and would lead to pernicious mechanisms of social control.

Similar arguments have been raised over any research that associates genetically mediated characteristics with human behavior. A major source of the conflict over recombinant DNA research is its potential for removing the obstacles to genetic engineering by allowing scientists to transfer hereditary characteristics from one strain to another. This evokes images of eugenics: a poster at a recombinant DNA forum read, "We will create a perfect race." Furthermore, the search for genetic information is suspected of being a means to justify and perpetuate existing social and economic inequalities.

Research on the relationship between race and IQ provokes similar visions of abuse. Do we really want to know if there is a genetic basis for intelligence or for social behavior? Would this not open the door to the development of mechanisms for social control? If IQ were related to heredity or race, could this result in social stigmatization which could itself have deleterious consequences? Moreover, do we really want to be able to control human qualities—to shape physical or mental characteristics?

Concern about misuse of scientific findings also contributed to the halting of research on the use of psychotropic drugs to treat children with learning disabilities. Scientists in Boston proposed to use as research subjects boys with functional behavioral disorders who were referred by their schools or by mental health clinics. The effect of the drugs on their behavior at home and at school, and on their academic achievement, was to be observed and measured against a control group of boys given placebos. Despite prior discussions with parents, with hospital and university review boards, and with various school and community groups, the project became extremely controversial. The major concern was the danger inherent in the use of a "medical fix" that could be used to control behavior under conditions that preclude rational discourse and independent decision making. Above all, the use of drugs to control behavior could divert attention from inequitable environmental sources of behavioral problems—poverty, poor housing, inadequate schools. In 1974, the Massachusetts state legislature prohibited the administration of psychotropic drugs to children for purposes of research.

Fear of the social misapplication of biomedical research is heightened by reports of the use of medical technologies to subdue aggressive inmates of penal or mental institutions and to treat hyperactive children. Anxiety about such misapplication leads directly to questions about the advisability of seeking certain kinds of knowledge at all; for, it is argued, knowledge cannot easily be disengaged from its uses. Biology, asserts a group of critical scientists, is a "social weapon."

Moral and Ethical Concerns

For many critics, it is the moral implications of science and technology that shape their dissent: the fear that science may change the normal state of nature, alter the essential genetic structure of man, or threaten cherished beliefs. At a time when the accomplishments of science have fostered in some a faith in rational explanations of nature, there are concerted efforts by others to reinvest educational systems with traditional faith. And even as biomedical research brings about dramatic improvements in medical care, there are always critics who seek to block research and to question areas of science that challenge traditional values.

Moral issues compound the recombinant DNA dispute. The most outspoken critics of this research are less concerned with its risks and benefits than with the creation and control of new life forms, a possibility which violates deeply held beliefs about free will and self-determination. They use anxiety-provoking images of Frankenstein and Faust: "It is no longer science fiction but its realization"; "scientists hold our genetic future in their hands." Indeed, this dispute has crystallized the persistent fears that scientific inquiry, as increasingly it approaches the study of man, flouts cherished ethical principles.

On similar grounds, fundamentalists take issue with the teaching of evolutionary biology in public schools. They feel that it intrudes on their religious beliefs and demand that schools provide "equal time" for creation theory. They are a sufficiently powerful force that many educators, biologists, and publishers view their demands as a serious threat to science education.

Concern about the moral implications of science has also nurtured the dispute over fetal research. Research on the human fetus is one of the more productive branches of medical science; it contributes to many improvements in medical practice, such as the decline in infant mortality and the development of new drugs and vaccines. Such research, however, has also been highly controversial, linked closely as it is to the issue of abortion. In the eyes of "right to life" groups, fetal researchers are using the "victims of abortion" as human guinea pigs. This is a moral affront, an offense to basic beliefs.

This dispute polarized over the definiton of the fetus itself. Is it human, with a full complement of rights, or is it simply a mass of tissue? Critics argued that the fetus is human and that respect for human dignity must extend to every level of biological development. Scientists, on the other hand, claimed the fetus to be simply a mass of tissue, an extension of the mother, with no independent moral or legal standing. Thus, they contend that the size, weight, and other technical characteristics of the fetus provide acceptable criteria for its use in research. While the critics felt that human dignity and the integrity of the individual militate against all such research, scientists argued that the medical knowledge to be obtained from fetal research is sufficient justification for work in this area.

The arguments against genetic manipulation and fetal research follow a long tradition of criticism of science as disruptive of human values, of nature, of natural law. Indeed, such an attitude accounts for the very emotional character of many recent conflicts over scientific research.

Questions of Equity

Questions of equity and justice in the allocation of the costs and benefits of science and technology are common sources of conflict. Given the high financial costs of research, how should priorities be determined? Should funds be allocated for research to develop sophisticated and costly life-saving technologies or for more basic health-care procedure? If a research technique promises social gain but poses individual risk (for example, research on human subjects), which should prevail? Conversely, if research will benefit only a select few, should the public pay the cost?

Many questions of equity originate with the rising cost of research: national expenditure for biomedical research grew from $110 million in 1950 to $2.5 billion in 1975. Although this is only a small percentage of the total health budget, the questioning of research funding and priorities reflects a broader concern with the health program as a whole. Considerations of cost and relevance have long affected decisions about technology; they are now increasingly applied to the allocation of funds for basic research as well.

The rapidly developing field of perinatal research, for example, raises a number of such questions. It can cost over $40,000 to save one premature infant, yet a major source of prematurity is poor nutrition and neglected health of pregnant mothers. Could these funds be more effectively used to ease the social and environmental problems that contribute to the incidence of prematurity? Is the research devoted to saving 700-gram infants an appropriate expenditure of public funds? And if there are limited resources, how can we establish criteria to decide who should be saved? Similar questions have been asked about the development of sophisticated medical technology for prolonging the life of comatose patients or saving people with special and costly medical needs.

Disputes over questions of equity commonly occur when citizens in a community become aware that they are to bear the costs of a technology intended to benefit a different or much broader constituency. This concern is most explicit in controversies over the siting of large-scale technologies. Airports and power plants serve large regions, but proximate neighbors bear the environmental and social burden. Normally, projects are planned and sites selected on the basis of economic efficiency and technical criteria. But community protests raise basic questions of distributive justice: can any reduction in some citizens' welfare be justified by greater advantages to others? Can the magnitude or intensity of costs borne by neighbors of a major project be reasonably incorporated into cost-benefit calculations?

Freedom of Choice and Individual Rights

The expansion of clinical research along with several scandals involving the abusive use of experimental subjects have raised concern about the threat to individual rights inherent in the use of human subjects for scientific experimentation. Rules established at the Nüremburg trials require the voluntary and informed consent of human subjects, but these ideals are often problematical to implement. Efforts to define "informed and voluntary consent" have raised difficult questions. What constitutes adequate information? Are many subjects subtly coerced to participate in experiments?

Difficulties are compounded by the disproportionate use of prisoners, the mentally retarded, the elderly, and other institutionalized groups as conveniently available subjects. Often, research subjects are hospitalized patients with rare or terminal illnesses, and it is thus impossible to differentiate therapy from research. Legislation has provided greater protection for the rights of vulnerable people, restricting research on those who are clearly unable to make informed judgments. The exposure of some people to harm in order to benefit society as a whole remains, however, a source of tension in the search for adequate guidelines to protect individual rights without constraining useful research.

Questions of individual rights also arise when decisions about the implementation or regulation of technologies preclude free choice. Some technologies, by definition, intrude on individual choice. The implementation of a process such as fluoridation or the installation of automobile air bags mandates universal compliance; everyone must partake of the decision and its consequences. In other cases, the potential harm resulting from technology encourages government controls. However, if the sale of saccharin, cyclamates, or laetrile is prohibited, individuals who want these products are denied the right to buy them; the government imposes regulations on the assumption that individual choices may have social costs, or that individuals may fail to make rational and enlightened decisions on their own behalf. Such constraints may also be seen as the protection of professional privilege, as unnecessary government paternalism, or as a violation of individual rights.

These are the views that have maintained the laetrile dispute.

These diverse concerns have brought decisions about science and technology into the realm of adversary politics. But whatever motivates individual disputes, the ubiquitous question in all controversies is "Who should control crucial policy choices?" Reflected in this question is a troubling feature of contemporary society—the impact of increased technical complexity and specialization on the decision-making process. Deference to technical knowledge enables those with specialized competence to exert a powerful influence on public agencies. The power given to those who control technical information can threaten democratic principles, reducing public control. The legitimacy of government, it is feared, may come to be based less on representation than on control over the context of facts and values in which policies are shaped. Thus, the role of expertise has revived the debate about technocracy. This has been a driving force behind controversies over science and technology, expressed in the recurrent populist theme of participation. Indeed, participation is increasing in importance as an ideology in American society even as technical complexity acts to limit effective political choice.

A Political Challenge

If there is conflict between scientific and technical interests and public concerns, does one rely on professionals to assess and control the impact of their work, or on those who may be affected? Assumptions about the importance of technical competence as the basis for legitimate decision-making authority are changing, as many lay groups claim social competence to participate in the evaluation of science.

Sources of Opposition

Many different grievances stimulate opposition to science and technology. Those who live in the vicinity of an airport, a nuclear-waste disposal facility, or a potentially hazardous research site have practical reasons to protest. Directly affected by land appropriation, noise, immediate risks, local economic or social disruption, or by some encroachment on their individual rights, they mobilize to influence policy directly through issue-oriented and elite-challenging groups.

There are also active critics of science and technology who are not directly affected by specific projects but who are motivated by more general ideological concerns. These include members of major environmental and public-interest associations and those who seek a more spontaneous or natural society.

And there are people who oppose science and technology because of global political concern. They see science as an instrument of military or economic domination and use the science policy arena as a means of advancing their political ends.

Conflicts over science and technology draw support from sharply contrasting social groups. The most active is the middle class: educated people with sufficient economic security and political skill to participate in decision-making. But conservative fundamentalists in California who are seeking to influence a science curriculum share concern about local control with citizens in a liberal university community who are seeking to influence the siting of a power plant.

Finally, an important source of criticism has been the scientific community itself. Many young scientists became politicized during the 1960s. At that time, they focused on antiwar activities, on university politics, and on the issue of military research in universities. More recently, their attention has turned to the environment, energy, biomedical research, and harmful industrial practices. These scientists often initiate controversies when they raise questions about potential risks in areas obscured from public knowledge.

The mobilization of opposition groups depends on a number of factors. If the decision in question (for example, the siting of a power plant) directly affects a neighborhood, local activists are relatively easy to organize on the basis of immediate interests. Many issues (such as recombinant DNA), however, have no natural constituency. Risks may be diffuse. The interests affected may be hard to define, or so dispersed as to be difficult to organize. Or the significant interests may be more concerned with employment than with their environment and therefore willing to accept certain risks. In such cases, participation is limited, and the controversies revolve mainly around technical debate among scientists.

The Political Role of Experts

Technical expertise is a crucial political resource in conflicts over science and technology, because access to knowledge and the resulting ability to question the data used to legitimize decisions are an essential basis of power and influence. The activist-scientist has played an especially important role in formulating, legitimizing, and supporting the concern that is expressed in recent controversies. The

willingness of scientists to expose the technical uncertainties that underlie many disputes, and to lend their expertise to citizen groups, constitutes a formidable political challenge.

However, protest groups also make extensive use of technical expertise to challenge policy decisions. Power plant opponents have their own scientists, who legitimize their concerns about thermal pollution. Environmentalists hire their own experts to question the technical feasibility studies of questionable projects. Laetrile supporters have their own medical professionals. Even fundamentalists, who seek to have the Biblical account of creation taught in the public schools, present themselves as scientists and claim that "creation theory" is a scientific alternative to "evolution theory."

Whatever political and ethical values motivate these controversies, the debates often focus on technical questions that require such expertise. A siting controversy may develop out of concern with the quality of life in a community, but the debate may well revolve around technical questions—the physical requirements for the facility, the accuracy of the predictions that established its need, or the precise extent of environmental risk. Concern about the freedom to select a cancer therapy tends to devolve into technical arguments about the efficacy of treatment. Moral opponents of fetal research engage in scientific debate about the precise point at which life begins.

This is tactically effective, for in all disputes broad areas of uncertainty are open to conflicting scientific interpretation. Decisions are often made within the context of limited knowledge about potential social or environmental impacts, and there is seldom conclusive evidence for reaching definitive resolution. Thus, power hinges on the ability to manipulate knowledge—to challenge the evidence presented to support particular policies—and technical expertise becomes a resource exploited by all parties to justify their political or economic views.

When expertise becomes available to both sides in a controversy, it further polarizes conflict by calling attention to areas of technical ambiguity and to the limited ability to predict and control risks. The very existence of conflicting technical interpretations generates political activity and leads directly to demands for a greater role for the public in technical decision-making. After hearing 120 scientists argue over nuclear safety, for example, the California State Legislature concluded that the issues were not, in the end, resolvable by expertise: "The questions involved require value judgments and the voter is no less equipped to make such judgments than the most brilliant Nobel laureate."

The Participatory Impulse

Conflict has brought science into the public arena, where it is subject to the pressures of adversary politics. The tactics by which various groups seek to influence decisions about science and technology range from routine political actions, like lobbying or intervention in public hearings, to litigation, referenda, and political demonstrations. These channels of participation depend on the institutional framework or political system in which opposition takes shape.

In the United States, litigation has become a major means for citizens to challenge technology. The role of the courts has expanded through the extension of the legal doctrine of standing—private citizens with no alleged personal economic grievances may bring suit as advocates of the public interest. Between 1970 and 1975, 332 suits were filed against federal agencies for inadequate environmental impact statements. In the Federal Republic of Germany, the courts have assumed a similarly important role, especially with regard to the nuclear debate. Several lower courts have, in fact, sustained citizen appeals by declaring moratoria on the construction of nuclear power plants.

The courts have also challenged scientific research practices, as in the litigation in Boston over fetal research. While there are few examples of long-term legal obstruction of research, the threat of litigation remains a powerful means of public influence.

The participatory impulse has also forced elected representatives, seeking to maintain popular support, to consider technical issues that are normally beyond their political jurisdiction. For example, the Cambridge (Massachusetts) City Council and local government bodies in other university communities claim the authority to judge the adequacy of safety regulations in biology laboratories. Similarly, many local townships forbid the transport and disposal of radioactive materials within their jurisdiction.

Concern about science and technology is also expressed in expanded Congressional activity. The *laissez-faire* attitude that has long characterized Congressional policy towards science is being replaced by greater public control. In April 1975 the U.S. House of Representatives passed an amendment to the National Science Foundation Authorization Bill requiring monthly Congressional review of all scientific research grants awarded by the Foundation. Although the bill, with all its cumbersome implications, was eventually defeated, its support in Congress jeopardized the autonomy of the scientific community. Worries have been reinforced by Senator William Proxmire's "golden fleece" awards given, with a flourish of publicity, to projects that he judges to be "frivolous." Several Congressmen have also investigated the peer review system, thereby questioning the very foundation of self-regulation in science. And Senator Edward Kennedy, a major Congressional leader in biomedical affairs, has been critical of the autonomy of science, calling for greater public participation in technical decisions.

The participatory impulse is also reflected in a tendency to place decisions about science and technology on the ballot for direct citizen vote. Referenda have been a prominent and growing feature of the political landscape and have been applied to such technical issues as fluoridation, nuclear power plant development, and airport siting.

Finally, in some cases local community action has taken the form of mass demonstrations, as in the protests against the landing of the Concorde and against the construction of airports and nuclear power projects. Even research practices have provoked public demonstrations: recall the antivivisection marches in protest against experiments on cats at the Museum of Natural History in New York.

Such actions suggest a changing relationship between science and the public. But demands for participation do not necessarily imply anti-science attitudes. More often they suggest a search for a more appropriate articulation between science and those affected by it. In its report on recombinant DNA research, the Cambridge Review Board thoughtfully expressed a prevailing view.

> Decisions regarding the appropriate course between the risks and benefits of potentially dangerous scientific inquiry must not be adjudicated within the inner circles of the scientific establishment. . . . We wish to express our sincere belief that a predominantly lay citizens' group can face a technical, scientific matter of general and deep public concern, educate itself appropriately to the task, and reach a fair decision.

This view presents science and technology policy as no different from other policy areas, subject to political evaluation that includes intense public debate. From this perspective, the protests are less against science and technology than against the power relationships associated with them; less against specific technological decisions than against the declining capacity of citizens to shape policies that affect their interests; less against science than against the use of scientific rationality to mask political choices. This is the political challenge posed by the disputes over science and technology, and it has profound implications for their resolution.

REVIEW QUESTIONS

1. Explain Jacqueline Feldman's meaning of the term *scientism*.

2. Why does Bernal (as quoted by Feldman) say that the learning of scientific method, today, is a farce?

3. Feldman refers to fragmentation in science. How does this "fragmentation principle" affect science and the scientist?

4. Throughout her article, and especially in the sections "An imperialist institution of knowledge," "Science against knowledge," and "Knowledge for the people," Feldman reveals a particular philosophical or political point of view. Had you detected it in your reading? What is the overall philosophical orientation of her paper? Does it render her comments invalid?

5. In her article "Science as a Source of Political Conflict," Dorothy Nelkin quotes former presidential adviser Zbigniew Brzezinski's classification of protests against such issues as recombinant DNA and nuclear power research as the "death rattle of the historically obsolete." What does Brzezinski mean? Do you think he is correct?

6. Describe the controversy over the definition of the fetus in scientific research.

7. Describe Nelkin's meaning, in the section "Questions of equity," when she refers to allocating funds to save premature infants. What is your opinion?

8. Can experts solve the problems addressed under "Questions of equity"?

9. What is the attitude of the Cambridge Review Board toward solving the ethical and moral problems of scientific research? *Give your own thoughtful answer to the problem.*

10. Do you believe that the lay person (nonscientist) has a responsibility to help decide issues which have complex scientific aspects, such as whether or not nuclear power plants are too dangerous to build or to rely on as a source of energy? Explain.

2 Asking Questions

Interaction of Organisms with Their Environment

About 14,000 years ago, somewhere in the "fertile crescent" of the Middle East, a man who lived in a cave asked himself a question. The question, translated from his primitive language to ours, began "What if . . .?" The rest of the question could have been ". . . I dig up some of the grass I have been gathering and plant it close to my cave?" Or he might have asked "What if I capture some of the goats I have been hunting and keep them penned up near my cave?" It is not clear exactly where each of these thoughts occurred, but evidence points to the area bounded on the west by the Mediterranean Sea and on the east by the Tigris and Euphrates rivers. There is evidence that by 8000 B.C. farming and animal husbandry had both been discovered there. What is most interesting is that although humans had been living in this area for at least 40,000 years, two of the most important discoveries ever made were made within a thousand years of each other.

1. Do you think that this was just a coincidence? _____ Explain. _____

_____

_____

_____

_____

_____

Do not read on until you have answered the question! Never leave a question unanswered. Since the exercises in your book are arranged progressively, you cannot answer the next question until you have answered the immediate one.

The discovery of agriculture led people to live close together near areas suitable for farming. Small settlements sprang up. A surplus of the "grass" which was cultivated (a form of wheat called "emmer") and meat from the flocks of goats made it possible for some members of the community to specialize in certain crafts and to trade their products for food. Potters appeared. Tradesmen traveled about seeking shells to decorate the body or obsidian (a black glass-like stone formed when sand is heated by volcanic action) to make tools with. There was at least one "city" with a large tower and walls, at Jericho (of Biblical fame), by 5000 B.C.

The men who asked themselves the questions which resulted in the discovery of agriculture brought about the "Neolithic Revolution," a period of intellectual ferment and social change rivaled by only a few periods in history, such as the Renaissance and our own era.

Evidence of the pattern of cultural evolution leading to the Neolithic Revolution is found in the stone tools uncovered in the caves at successively greater depths. Each layer of earth represents a different period of time, the earlier periods being found in the deeper layers. A succession in the evolution of tools is shown in Figure 2–1. All drawings are full-size.

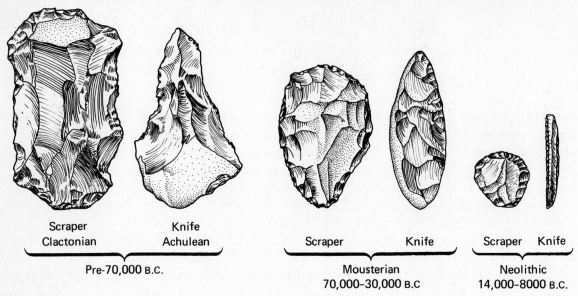

| Scraper | Knife | | Scraper | Knife | | Scraper | Knife |
| Clactonian | Achulean | | | | | | |

Pre-70,000 B.C.   Mousterian   Neolithic
                                   70,000–30,000 B.C   14,000–8000 B.C.

FIGURE 2–1   The evolution of flint (stone) implements over three successive periods of human cultural evolution.

2. What two factors do you notice about the tools as time progresses toward the present?

a. _____

_____

b. _____

_____

The small blades (microliths) of the Neolithic period were actually mounted in wooden or bone handles to form compound tools, such as the sickle shown in Figure 2–2.

Flint blades embedded in wood or bone handle

FIGURE 2–2   A Neolithic sickle.

The preceding discussion was designed to show you two things: (1) that the asking of questions about the environment had led to profound changes in man's cultural development and (2) that these questions are not isolated from, but instead are inspired by, the environment. In other words, man's capacity to pick out aspects of the environment and examine them, and then to ask himself questions about them, has led to our enormously complex and sophisticated society.

Other animals—dogs and monkeys, for example—have not been able to inquire about their surroundings. Hundreds of thousands of years ago they acquired patterns of behavior which they still employ today, virtually unchanged. Moreover there is a group of humans living in a "land of plenty" called Tierra del Fuego at the southern end of Chile. Climate and other environmental conditions have made it possible for these people to pluck food from the bough or the sea. As a consequence of having all their needs fulfilled easily, the Tierra del Fuegans have apparently had no need to ask questions. No questions, no culture. Practically no language. Virtually no agriculture and few social institutions. It appears to be no accident that those early cultures which impress us with their creativity are found in areas physically challenging to man. The fabulous Minoan civilization in ancient Crete and the culture of Greece, the pinnacle of all ancient cultures, sprang out of rocky mountainous environments.

Modern man has conquered his physical environment. We live in "climate-controlled" homes and we work in air-conditioned offices and schools.

3. Does this mean that we are placing ourselves in the position of the natives of Tierra del

Fuego?_____Explain in detail in the space below.

4. Which aspects of our environment (environment may be defined as *all* aspects of the world around us) stimulate your curiosity? Write some questions which need answering, questions arising from those aspects of our environment which you have just described.

5. In what direction do you think human cultural evolution is going? (That is, what will our society and environment be like in two hundred years?)

The discoveries of animal husbandry and farming occurred relatively close together, the discovery of antibiotics occurred simultaneously in two parts of the world, and the two vaccines against poliomyelitis were created within a year of each other.

6. What do these facts suggest about *when* questions are asked? _____

_____

_____

_____

7. If you are not sure of your answer to the above question, consider the following: What leads to the creation of a question? Is a question an isolated thought which springs into your

head in a moment of inspiration?_____

_____

_____

_____

The purpose of this exercise is to impress you with the importance of questioning the things which are going on around you. The more inquisitive you are and the less you consider yourself a passive receptacle of someone else's knowledge, the better you will function in the spirit of science.

PRELIMINARY INFORMATION

During today's laboratory exercise you will be presented with a miniature "world." To the animals and plants inside the container, the world is literally circumscribed by its walls. Yet virtually all the conditions to which the organisms were exposed in the wild still hold in their new environment. There are still the eaters and the eaten. There are still the pressures of the **physical environment** wherein some organisms seek light, and others dark; some respond to small changes in temperature, while others are less sensitive. The **biological environment** still presents challenges to survival. Availability of food, the danger of disease, and predation are ever-present in the small world on your laboratory table.

Your role is to explore this world. When you concentrate on the interactions between the organisms, this miniature world is as interesting as the moon or the top of a mountain. A commonplace snail becomes a puzzle. What does it eat? Where does it prefer to live? Exactly how does it move? Can it see? Can it hear? Can it taste? Can it feel?

Let your imagination flow freely. Try to understand what is happening in the container. *Ask questions.* Ask your neighbors, ask yourself; **inquire of the inhabitants of the container.**

■ THE PROBLEM: *What are some of the interactions among the organisms and between the organisms and their environment?*

Work in groups of four. List below as many questions as you can concerning both individual organisms and their interactions with one another. Number each question, beginning with 1.

Examine the questions. Can you put them into categories? Arrange them in groups below. Put the number of the question next to the category it belongs to. For example, a category might be "Questions about plants." Question 1, 7, 14 might go under this heading. Put your questions into categories now. Then examine your categories. Are there any obvious *kinds* of questions you might add to your list?

One category you may have neglected is "Questions solvable by using available equipment," such as "Will snails always crawl upward if placed in a test tube which is inverted every five minutes?" List those questions which you feel you can find the answers to today, using relatively simple equipment which is either visible in the room or the instructor is likely to have.

Choose the question which you and the other members of your group agree is the most interesting and meaningful, yet susceptible to solution by means of available equipment. *Simple questions are most readily solvable.* Complex questions, while they may appear likely to elicit more important answers, often yield no information because they are unanswerable. If you are interested in uncovering what seems to be a relatively complicated piece of information, try to break it down by asking a sequence of simple questions which, together, will contribute toward the final answer.

Choose your question now and write it under *Problem*.

Problem

What do you think the answer will be? Write this under *Hypothesis*.

Hypothesis

How do you plan to answer the question? Write your plan of operations under *Protocol*.

Protocol

How will you make sure that the result you observe is the answer to the question and not just a chance occurrence? Describe your means of proving that your data are not the result of chance under *Control*.

Control

What data will you uncover? When you formulate the plan for your investigation, try to build in a method of obtaining clear-cut data, preferably in numerical terms as, "Eight snails responded positively. Two responded negatively." Place your data, together with an explanation of their meaning, under *Results*. After you have performed your investigation, you will have, it is hoped, an answer to your question. Place the answer under *Conclusions*.

BEGIN YOUR INVESTIGATION NOW.

Results

Conclusions

3 A Test of Aspects of Scientific Thinking

This test has been devised to measure your ability to think scientifically. It is divided into several parts, each part testing a different phase of scientific thinking.

General Directions

(1) Answer all items: Mark those you are unsure of so that you will be able to come back to them during the classroom discussion period. The instructor will not tell you whether or not your answers are correct. You will have the opportunity, however, to work out the correct answers with the members of your group. At the end of the semester you will be graded on your performance on a test which will have a format similar to this one.

(2) Circle the appropriate number on your answer sheet.

(3) Each item has only one answer. Select the best answer and circle no more than one number for each question.

Instructions

This portion of the test is designed to measure your ability to differentiate phases of thinking. These steps include major problems or perplexities, possible solutions to problems, observations which are not results of experimentation but rather preliminary observations, results of experimentation, and finally conclusions.

The following key is to be used for the succeeding paragraph. Certain parts of the paragraph are underlined, and each underlined item is to be considered a question. Choose the proper response from the key and circle the appropriate number on the Answer Sheet (page 41).

Key

(1) A major problem (stated or implied).
(2) Hypothesis (possible solution to problem).
(3) Result of experimentation.
(4) Initial observation (not experimental).
(5) Conclusion (probable solution to problem).

(1) How a homing pigeon navigates over territory it has never seen before is not fully understood. (2) Do air currents stimulate the pigeon in some way? (3) Are pigeons equipped with some sort of magnetic compasses; that is, are they sensitive to the earth's magnetism? A scientist tested the latter by fastening small magnets to the wings of well-trained pigeons. (4) Most of these birds never got home. (5) Others, carrying equal wing weights of

*This test was constructed by Dr. Mary Alice Burmester, Office of Evaluation Services, Michigan State University. It is reproduced with her kind permission.

nonmagnetic copper, made the home roost without trouble, (6) indicating that the earth's magnetism is a factor in pigeon navigation. However, a magnetic compass could not, by itself, bring the pigeon back to his roost, because many places on the earth's surface have identical magnetic conditions. The scientist attempted (7) to determine the other guiding factor. (8) It might be the sun or stars, but pigeons navigate under clouds. If pigeons were sensitive to some factor connected with the lines of latitude, they would have all they need to find their way home. The next step was (9) to find some physical force, something the pigeons might be able to detect, related to the lines of latitude. The effect of the earth's turning varies directly with latitude; objects near the equator are carried daily around the earth's circumference, moving at over 1000 mph. Objects near the poles are carried around more slowly. The direction and variation of this circling can be recorded by various man-made instruments. (10) Why shouldn't the pigeons feel it, too? If they could, they would have, along with their magnetic compass, a satisfactory navigating instrument. The investigator trained hundreds of pigeons to return to their home roosts. Then he took them to a place where the lines representing the earth's magnetism cross the parallels of latitude at the same angle as at their home roost. When he released the pigeons to the east of this spot the pigeons flew west. He believes that (11) pigeons are guided by both the earth's magnetism and by its turning. (12) Just where the birds' instruments for navigation are located is still unknown, but the scientist found that (13) birds have a mysterious organ in their eyes, at the end of the optic nerve. (14) This organ may contain the nerve fibers that pick up vibrations of magnetism and the sense that measures the earth's turning.

Instructions

This portion of the test is designed to measure your understanding of the relation of facts to the solution of a problem. The overall problem is presented. This is followed by a series of possible solutions to the problem (hypotheses). After each hypothesis there are a number of items, all of which are true statements of fact. Determine how the statement is related to the hypothesis and mark each statement according to the key which follows the hypothesis.

■ THE PROBLEM: *What factors are involved in the transmission of infantile paralysis?*

□ Hypothesis I: *Healthy persons having had contact with diseased individuals may carry the disease from one person to another.*

For items 15 through 22 circle number:

(1)   if the item offers direct evidence in support of the hypothesis.
(2)   if the item offers indirect evidence in support of the hypothesis.
(3)   if the item offers evidence which has no bearing on the hypothesis.
(4)   if the item offers indirect evidence against the hypothesis.
(5)   if the item offers direct evidence against the hypothesis.

15. Monkeys free of the disease almost never catch infantile paralysis from infected monkeys.

16. It has been found that exertion prior to or at the time of infection increases the incidence of the disease.

17. Even during epidemics cases are spotty; it is usually impossible to trace one case from another.

18. The virus is always found in the stools of people who have the disease.

19. Most persons in contact with the diseased individual do not develop the disease.

20. Nine out of fourteen adult contacts had virus in stools; almost all child contacts had virus in stools.

21. Up to two months after contact the virus is found in the stools of persons who contacted the victims, but who did not contract the disease.

22. In the stools of noncontacts the virus was found in only one person in ten.

23. What is the status of hypothesis I?

    (1) The hypothesis is true.
    (2) It is probably true.
    (3) The data are contradictory, so the truth or falsity cannot be judged.
    (4) It is probably false.
    (5) It is definitely false.

□ Hypothesis II: *The higher the degree of sanitation the greater are the chances of epidemic forms of the disease.*

For items 24 through 30 circle number:

(1) if the item offers direct evidence in support of the hypothesis.
(2) if the item offers indirect evidence in support of the hypothesis.
(3) if the item offers evidence which has no bearing on the hypothesis.
(4) if the item offers indirect evidence against the hypothesis.
(5) if the item offers direct evidence against the hypothesis.

24. Monkeys free of the disease almost never catch infantile paralysis from infected monkeys.

25. In India epidemics seldom occur.

26. In India children under five are about the only ones affected.

27. During the war there was one epidemic among the European and American soldiers in India; the incidence among the soldiers was extremely high.

28. The percent of cases of infantile paralysis in whites is about four times that in black people.

29. The percent of cases of infantile paralysis is higher in rural districts than in the cities.

30. During an epidemic nonparalytic cases outnumber paralytic cases ten to one.

31. What is the status of hypothesis II?

    (1) The hypothesis is true.
    (2) It is probably true.
    (3) The data are contradictory, so the truth or falsity cannot be judged.
    (4) It is probably false.
    (5) It is definitely false.

Instructions

This portion of the test is designed to measure your ability to interpret data and to test your understanding of experimentation. The numbers in the first column of the table below are the numbers which you will use as your answer. Thus the table is both the source of data and your key for the questions which follow it. In each case where a test tube number or group number is called for, the one which gives positive evidence for the statement should be given.

Where the control or comparison is called for, give the number of the test tube or group which offers a comparison.

For example:

(1) Leaf in dark—no starch.
(2) Leaf in light—starch.

"Light is necessary for the production of starch." You would circle 2 because this is the positive evidence, but it would be meaningless if it were not compared with the leaf in the dark. Therefore, the answer for the question "What is the control (comparison) for item 1?" would be marked 1.

Items **32** through **46** refer to the data presented below. Some test tubes were set up and each contained 1 gram of fat. They were marked 1, 2, 3, 4, and 5. Mark each item according to the test tube number called for. Various substances were added to the tubes containing fat. All substances were dissolved in water before they were added to the fat. All test tubes were kept at 30°C (water boils at 100°C). For test tube 5, substance A was boiled and then allowed to cool before it was added to the fat.

Test Tube Number	Initial Contents of Tubes	Amount of Substance B Present after 24 Hours
1	Fat + substance A	0.1 gram
2	Fat + substance A + substance C	0.5 gram
3	Fat + water	0.0 gram
4	Fat + substance C	0.0 gram
5	Fat + substance A (boiled)	0.0 gram

32. Give the number of the test tube which acts as a control (comparison) for the entire experiment.

33. Give the number of the tube which gives evidence that fat does not break down spontaneously into substance B in 24 hours.

34. Give the number of the tube used to show that a temperature of 30°C was not sufficient to cause fat to be broken down into substance B.

35. Give the number of the test tube which gives evidence that substance A is the active substance in the breakdown of fat to substance B.

36. Give the number of the test tube which is the control (comparison) for item **35**.

37. Give the number of the tube which provides evidence that substance C alone is ineffective in the breakdown of fats.

38. What is the control for item **37**?

39. Which test tube gives evidence that substance C accelerates the rate of activity of substance A?

40. Give the number of the tube which is the control for item **39**.

41. Which tube gives evidence that substance A is a substance whose properties can be destroyed?

42. Which is the control for the tube in item **41**?

43. Which tube gives evidence that substance C affects the fat in some way so that substance A can more easily act upon it?

44. Which tube is the control for **43**?

45. Which tube gives evidence that substance A is not a stable substance?

46. What is the control for item **45**?

Instructions

This portion of the test is designed to measure your ability to make conclusions. When facts are analyzed and studied, they sometimes yield evidence which helps in the solution of a problem. However, any conclusion must be checked before it can be accepted. The following key includes four ways in which conclusions may be faulty. Each of the items presents a question or problem, a brief description of an experiment, and one or more conclusions drawn from the experiment. Each experiment was repeated many times. Read each problem, experiment, and the conclusions. Where several conclusions are given, evaluate each conclusion separately. Is the conclusion tentatively justified by the data? If so, circle 1 on your answer sheet. If the conclusion is not justified, determine whether 2, 3, 4, or 5 in the key is the best reason for its being faulty and circle the proper number on your answer sheet.

Key
- (1) The conclusion is tentatively *justified*.
- (2) The conclusion is unjustified because *it does not* answer the problem.
- (3) The conclusion is unjustified because *the experiment* lacks a control comparison.
- (4) The conclusion is unjustified because *the data are faulty or inadequate, though a control was included*.
- (5) The conclusion is unjustified because *it is contradicted by the data*.

□ **Problem I:** *A person wanted to determine whether bile aided in the digestion of fat.*

He found that whenever he mixed pancreatic juice with fat a small part of the fat was digested. But whenever he mixed pancreatic juice and bile with fat, he found that the fat was completely digested. When he mixed bile alone with fat, he found that there was no digestion.

47. He concluded that bile aided in the digestion of fat.
48. Someone else claimed that bile does not aid in the digestion of fat.

□ **Problem II:** *To determine some of the requirements for the sprouting of seeds.*

Two groups of plants were planted in flowerpots. Conditions of both were the same except that one pot was put in the greenhouse at 5°C; the other group was put in a greenhouse at 21°C. Those in the cold room did not sprout; those in the warm room sprouted. Many kinds of seeds were used in each group.

49. Conclusion: A temperature of 21°C is required for seeds to sprout.
50. Another conclusion: Moisture is one of the requirements for the sprouting of seeds.
51. Another conclusion: Energy is needed for anything to grow.

□ **Problem III:** *An individual wished to determine whether oxygen is used during sleep.*

He analyzed the expired air of a large number of sleeping persons. He found that the expired air contained oxygen.

52. He concluded that oxygen is not used during sleep.
53. Another concluded that oxygen is needed for life.

54. Someone else claimed that people breathe while they are sleeping.

☐ **Problem IV:** *An investigator wanted to determine whether light increased the rate of a certain reaction.*

On repeated tests it was found that a certain amount of the original substance (X), after 1 hour, would produce 1 gram of substance Y with 10 photons (units of light) of illumination, 2 grams with 20 photons, 4 grams with 30 photons, and 3 grams with 40 photons.

55. Conclusion: Increased amount of light increases the rate of reaction.
56. Another conclusion: Heat increased the rate of reaction.

☐ **Problem V:** *A person wanted to know what caused a certain disease.*

He examined 1000 patients with the disease. All had a certain kind of bacteria (bacteria A) in the digestive tract.

57. He concluded that bacteria A caused the disease.

☐ **Problem VI:** *A person wanted to know why plants bend toward the light.*

He placed one group of plants in the light with the light source at the right. He placed another group of similar plants in the dark. The plants in the dark grew straight; the plants in the light were bent to the right.

58. He concluded that plants bend toward the light.

Instructions

This portion of the test is designed to measure your ability to interpret data. Following the data you will find a number of statements. You are to assume that the data as presented are true. Evaluate each statement according to the following key and mark the appropriate number on your answer sheet.

Key

  (1) True: The data alone are sufficient to show that the statement is true.
  (2) Probably true: The data indicate that the statement is probably true, that it is logical on the basis of the data, but the data are not sufficient to say that it is definitely true.
  (3) Insufficient evidence: There are no data to indicate whether there is any degree of truth or falsity in the statement.
  (4) The data indicate that the statement is probably false; that is, it is not logical on the basis of the data, but the data are not sufficient to say that it is definitely false.
  (5) False: The data alone are sufficient to show that the statement is false.

Items **59** through **76** refer to Figure 3–1. Use the key above to answer the items. The lizard is considered to be cold-blooded, the others warm-blooded.

59. When the external temperature is 50°C, the temperature of the lizard is also 50°C.
60. At an external temperature of 50°C, the temperature of the cat is 50°C.
61. When the external temperature is 50°C, the temperature of the anteater would be higher than the temperature of the cat.

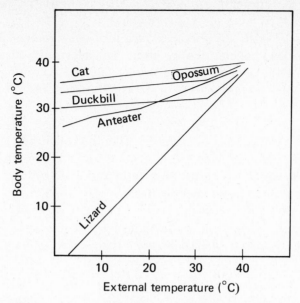

FIGURE 3-1 Relationship between external temperature and
the internal temperature of selected vertebrate animals.

62. The temperature of a mouse would be about halfway between that of the cat and the anteater.

63. At no time during the experiment did any of the animals have the same body temperature.

64. There is a close correlation between the body temperature of the lizard and that of the external environment.

65. At 20° below 0°C, the lizard would be frozen.

66. If the temperature of other cold-blooded animals were plotted, it would resemble that of the lizard.

Instructions

Items **67** through **70** are a re-evaluation of some of the preceding items. Reread items **59, 61, 64,** and **66** and determine whether each is a generalization, an extension of the data, an explanation of the data, merely a restatement of the data, or none of these. Answer each according to the following key.

Key

(1) A generalization; that is, when the data say it is true for this situation, a generalization says it is true for all similar situations.

(2) The data indicate a trend which if continued in either direction would make the statement true.

(3) An explanation of the data in terms of cause and effect.

(4) A restatement of results.

(5) None of the above.

67. Item **59.**
68. Item **61.**
69. Item **64.**
70. Item **66.**

Instructions

Items **71** through **76** are designed to measure your understanding of assumptions underlying conclusions. Two conclusions are given (not necessarily justified by the data). The statements which follow the conclusions are to be evaluated according to the following key. These items all relate to the data presented for items **59** through **66**.

Key
(1)   An assumption which must be made to make the conclusion valid (true).
(2)   An assumption which if made would make the conclusion false.
(3)   An assumption which has no relation to the validity (truth) of the conclusion.
(4)   Not an assumption; a restatement of fact.
(5)   Not an assumption; a conclusion.

□ Conclusion I: *Warm-blooded animals have some type of heat-regulating mechanism.*

71. The cat and the duckbill are very different in their reaction to the external environment.
72. The opossum had a lower body temperature than the cat.

□ Conclusion II: *Anteaters and duckbills are more closely related than anteaters and cats.*

73.   Similarity of reaction of living things indicates a relationship.
74.   The temperature of the anteater varied more with the external temperature than did that of the cat.
75.   The degree of closeness of similarity of response of living things runs parallel with the closeness of kinship.
76.   The temperature of the cat varied less than that of the anteater and duckbill with change of temperature.

ANSWER SHEET

Name _____

Instructor's name _____

1.	1	2	3	4	5	**39.**	1	2	3	4	5
2.	1	2	3	4	5	**40.**	1	2	3	4	5
3.	1	2	3	4	5	**41.**	1	2	3	4	5
4.	1	2	3	4	5	**42.**	1	2	3	4	5
5.	1	2	3	4	5	**43.**	1	2	3	4	5
6.	1	2	3	4	5	**44.**	1	2	3	4	5
7.	1	2	3	4	5	**45.**	1	2	3	4	5
8.	1	2	3	4	5	**46.**	1	2	3	4	5
9.	1	2	3	4	5	**47.**	1	2	3	4	5
10.	1	2	3	4	5	**48.**	1	2	3	4	5
11.	1	2	3	4	5	**49.**	1	2	3	4	5
12.	1	2	3	4	5	**50.**	1	2	3	4	5
13.	1	2	3	4	5	**51.**	1	2	3	4	5
14.	1	2	3	4	5	**52.**	1	2	3	4	5
15.	1	2	3	4	5	**53.**	1	2	3	4	5
16.	1	2	3	4	5	**54.**	1	2	3	4	5
17.	1	2	3	4	5	**55.**	1	2	3	4	5
18.	1	2	3	4	5	**56.**	1	2	3	4	5
19.	1	2	3	4	5	**57.**	1	2	3	4	5
20.	1	2	3	4	5	**58.**	1	2	3	4	5
21.	1	2	3	4	5	**59.**	1	2	3	4	5
22.	1	2	3	4	5	**60.**	1	2	3	4	5
23.	1	2	3	4	5	**61.**	1	2	3	4	5
24.	1	2	3	4	5	**62.**	1	2	3	4	5
25.	1	2	3	4	5	**63.**	1	2	3	4	5
26.	1	2	3	4	5	**64.**	1	2	3	4	5
27.	1	2	3	4	5	**65.**	1	2	3	4	5
28.	1	2	3	4	5	**66.**	1	2	3	4	5
29.	1	2	3	4	5	**67.**	1	2	3	4	5
30.	1	2	3	4	5	**68.**	1	2	3	4	5
31.	1	2	3	4	5	**69.**	1	2	3	4	5
32.	1	2	3	4	5	**70.**	1	2	3	4	5
33.	1	2	3	4	5	**71.**	1	2	3	4	5
34.	1	2	3	4	5	**72.**	1	2	3	4	5
35.	1	2	3	4	5	**73.**	1	2	3	4	5
36.	1	2	3	4	5	**74.**	1	2	3	4	5
37.	1	2	3	4	5	**75.**	1	2	3	4	5
38.	1	2	3	4	5	**76.**	1	2	3	4	5

II Morphology and Systematics

4 Inductive and Deductive Reasoning

Classification of Living Things: Form and Structure 1

You have learned that science is a logical method of considering reality. Since we deal with reality all of our waking lives, science is important as a means of ordering and making more logical our reactions to our environment.

The basic method of science is generalization, or **induction**. This is the process of drawing inferences about a whole class from a few of its members or making a generalization by tying together various related facts.

When a botanist describes a new plant, he may be primarily interested not in the properties of the individual specimens he observed, but in those properties he has reason to believe are shared by all other specimens of the same species. Thus a brown spot on one plant but not on the others would not be considered significant, but the color of the petals and the shapes of the flowers and leaves would be used to make up a generalized description of all plants of that species.

A composite of a particular length of dress, a particular line, and a particular collar might together describe a new style in women's clothing. To describe a fabric or a button would not be adequate to bring to mind a certain mode of dress, but the combination of several pertinent characteristics would serve as a general description.

The method of reasoning that builds facts into a generalization is called **induction** or **inductive reasoning**.

Deduction, on the other hand, proceeds from a known generalization and allows the prediction of some hitherto unknown facts or events. Deduction requires that a generalization has already been established. Knowing that flowers are characteristically found on all plants which have true roots, stems, and leaves, and bear covered seeds, would lead us to predict that a newly discovered plant with those characteristics would also have flowers—even if none are visible on the specimens shown to us. We have deduced that these plants must have flowers.

Sherlock Holmes, the great fictional detective, was always able to get his man by predicting the villain's actions on the basis of his huge store of "laws of human behavior," which were nothing more than generalizations about the way people behave.

Ordinarily it is not possible to examine every member of a large group of objects or ideas. You cannot examine every suit in the store, every corpuscle in the body, or every insect with sucking mouthparts. Consequently, induction is ordinarily based on a *part* of the class membership. This leads to the problem of selecting a representative sample from the group, a subject which will be discussed later on.

Scientific method is usually characterized by

(1) *Observation* of random unrelated facts.
(2) *Formulation of hypotheses* tying together some of these facts (*inductive reasoning*).
(3) *Validation of the hypotheses* in order to establish a generalization (theory, law).
(4) *Prediction* of some occurrence or previously undiscovered relationship among facts on the basis of the established generalization (*deductive reasoning*).
(5) *Testing the validity of the prediction* by further experimentation.

Note: You should not misconstrue the summary of intellectual processes presented above as a formula which, if consciously followed in a machine-like fashion, will yield the answer to any problem. The list simply defines the usual thought processes which scientists use when they solve problems. This kind of artificial dissection of the thinking process is a pedagogical device to show more clearly how facts are woven into ideas. There is no pat "scientific method" which can be taught abstractly. Each problem requires its own series of thought processes which will provide the key to its solution. For example, when you are mulling over a problem in your mind, you sometimes come upon the answer in a sudden flash of insight and cry "I've got it!" In psychological jargon, this is called an "inference leap." It should be clear that no matter how much you discuss this term on an abstract level, you have no guarantee that the necessary integration of ideas will appear again when you attempt the solution of another problem. On the other hand, if you realize that your inference was really just the careful arrangement of facts into a pattern in your mind, and not a magical process which defies understanding, you will be more inclined to attack your next problem in a logical fashion.

Scientific method is a general term for logical thinking based on fact. *It is not a specific list of processes to be followed mechanically.*

Summary: Induction is the process of looking at nature, picking out facts which are related, and placing them into a generalization. Deduction is using the newly acquired overview provided by the generalizations to make predictions.

1. After each of the following statements place a (D) if deductive or an (I) if inductive, and explain the reason for your choice.

    a. Chemists state that salt crystals are white. _____

_____

    b. A person looks at his watch and begins to run toward a train station even though

he does not hear or see a train coming. _____

_____

    c. A person has bought the products of a certain company and has repeatedly found them to be defective. He refuses to buy products made by that company any more.

_____

_____

    d. A football fan watches the performance of the various teams and picks out the

team he thinks is best. _____

_____

e. The fan bets that his team will win the next game. _____

_____

f. A fisherman looks at the water level of the stream, watches to see what the fishes

are feeding on, and then chooses a lure to fish with. _____

_____

g. A pioneer in "Indian country" sees clouds of dust on the horizon and, based on

past experience, looks for a place to hide. _____

_____

Note: Some of the above statements may be inductions or deductions, depending on your interpretation. Your explanation will reveal the logic of your choice.

PRELIMINARY INFORMATION

In the last centuries before our own, people exhibited an enhanced thirst for knowledge. Since there was relatively little known about the world, this frenetic search for an understanding of their surroundings concentrated on extracting generalizations—seeing relationships between organisms and processes that were hitherto undiscovered. The pace quickened even more in the twentieth century, for modern scientists have, in some fields, reached an "age of deduction," based on the inductions of their predecessors.

One area of study was concerned with making sense of the confusion of animals and plants which had been previously described. In antiquity Theophrastus reported on some 500 plant species. By the 18th century Linnaeus alone classified more than 18,000 plants. It became evident that this material would have to be organized. Linnaeus developed the system of binomial nomenclature (giving animals and plants two names (bi = 2; nomial = names), so that a robin could be called *Turdus migratorius* and a wolf, *Canis lupus*. The first name, the genus, is the broader of the two, like Jones in a person's name; the second, the species, is the most specific, like Arthur or Mary. Thus there can be several dog-like animals in the genus *Canis* (*Canis familiaris*, the common dog; *Canis latrans*, the coyote; etc.) The species tells us which dog-like animal is being specifically referred to. Thus a scientific name is similar to Jones, Mary; Jones, Arthur; or Jones, Bessie, in our own system of binomial nomenclature. But dealing with animals on a first name basis is for advanced classes. We will confine courselves to finding out what fundamental groups of animals exist on the earth. These groups are kingdom, phylum, class, order, and family. The **kingdom** is the broadest category. In the past all life was easily divisible into two kingdoms: plants and animals (see Chapter 5 for the modern concept of kingdom). The **phylum** divides a kingdom into major groups; such as worm-like animals or animals with soft bodies and shells. The **class** subdivides each phylum. For example, the phylum containing animals with soft bodies and shells can be further divided into classes containing soft-bodied animals with one shell (snails), with two shells (bivalves such as clams), and so on. Orders and families are successively more precise, and they are followed by genus and finally species, the actual name of the organism.

The following classification of a human will serve to summarize the above.

Phylum Chordata
Subphylum Vertebrata
Class Mammalia
Order Primates
Family Hominidae
Genus *Homo*
Species *sapiens*

Let us draw a parallel between biology and our everyday lives. When you go to a clothing store, it is often divided up into two major areas—men's clothes and women's clothes. You may then find sections entitled "sports clothes," "formal clothes," "business clothes," etc. Under sports clothes you might find a category "jackets" and under this heading may be a subdivision called "blazers." Instead of wandering around the store, you can proceed directly to the "jackets" section, part of which will be devoted to blazers. Your selection, simplified by the aforementioned organization, may fit into the category "two-button blue blazer," which does not vary much in purpose from a more biological category like Yellow-bellied Sapsucker (a bird) or *Ursus horribilis* (the Latin name for the grizzly bear).

There are huge collections of preserved organisms in museums around the world. By observing their similarities taxonomists have evolved criteria for placing them into groups. These characteristics form the basis for our modern system of classification and are the backbone of evolutionary theory. Until relatively recently, all living things were placed into two major groups. What are they? _____ and _____ . At this point in time our expanding knowledge has caused us to expand these original groups from two into five major divisions. These will be covered in succeeding laboratory sessions. For the time being, you will have your hands full trying to make even a few generalizations from the many specimens strewn about the room, so we will keep the two kingdom system for this period. At your table you will find bottles of preserved organisms. Your job will be to place them in groups according to their common characteristics, creating your own system of classification by induction.

■ THE PROBLEM: *To arrange organisms of various types in a system of classfication which demonstrates relationships between them.*

Follow the format outlined below, starting with the broadest groups and gradually becoming more specific. On each line describe the major distinguishing characteristics of a group. Include as many of the distinguishing characteristics which are common to the group as you can.

Level I (Kingdom): What are the two major kinds of living things? Give their names and their distinguishing characteristics on the lines below.

Name	*Distinguishing Characteristics*
1. _____	_____
2. _____	_____

Level II (Phylum): Classify the *animals* into groups containing similar types, for example, jointed-legged animals with hard outer coverings (exoskeletons), animals with shells and soft bodies. Do not name the groups; just give their characteristics.

1. <u>Jointed-legged animals with hard outer coverings</u> _____

2. <u>Soft-bodied animals with shells</u> _____

3. _____

4. _____

5. _____

6. _____

7. _____

8. _____

9. _____

10. _____

11. _____

12. _____

Level II (Class): Divide into separate sections the Level II subgroups mentioned below.

Jointed-legged animals with *exoskeletons* (shell-like, hard covering).

1. _____

2. _____

3. _____

Soft-bodied animals with *shells.*

1. _____

2. _____

3. _____

Another group (give characteristics of class). _____

_____

1. _____

2. _____

3. _____

Some people have attempted to place organisms into groups based on functional similarities rather than physical traits. For example, it has been suggested that all "swimming animals" be placed into one group and all "flying animals" be placed in another.

2. Suppose the main criterion for one of the phyla you established was "Those animals which have wings and can fly." List as many types of flying animals as you can.

_____

_____

_____

_____

_____

3. Look at as many of these kinds of animals as are available in the room. Describe below the pros and cons of placing them together.

4. Do you still believe that they all belong to one phylum? Why or why not? _____

_____

_____

5. Which type of reasoning was employed in this laboratory exercise, induction or deduction? _____ . Explain _____

_____

_____

6. Once the information has been organized into major groupings, what value might it have for future biologists who have access to your work? _____

_____

_____

5 Deduction I

Classification of Living Things: Form and Structure 2: Animals

The preceding laboratory exercise demonstrated that the first step in the process of problem solving is to accumulate information about the nature of the problem. This information is then organized in a manner which facilitates the solution of the problem by interrelating the information (tying it together). You will recall that the process by which random bits of information are drawn together into a generalization is called **induction** and that the use of the generalization to predict new relationships is called **deduction**.

Unrelated facts cannot be useful until they are brought together into a conceptual whole. One can learn little about the patterns and principles of the morphology (body structure) of organisms by simply looking at each separate plant or animal without attempting to relate it to the others. You have been given the opportunity to find some of the natural groupings into which animals fit, according to their morphological similarities. It should be understood that your inductions were based on the study of only a few representatives of the animal kingdom, and the criteria for your groupings were for that reason necessarily superficial.

PRELIMINARY INFORMATION

There are huge collections of preserved animals and plants in museums around the world. By studying this panorama of organisms, zoologists and botanists have evolved criteria for their classification. Although there is an almost infinite diversity in the shapes of living things, certain physical characteristics common to large groups of animals and plants become apparent when they are seen together. These characteristics form the basis for our modern system of classification and are the backbone of evolutionary theory. Some of the important criteria for classification are discussed below.

Symmetry

Most animals are either **bilaterally** or **radially symmetrical.** We can take advantage of this natural division in structure to separate these animals into two groups called **Bilateria** and **Radiata**.

The Bilateria are characterized by the fact that there is only one plane on which you can cut them in two and obtain two mirror images. Thus a dog can be divided into identical (but reversed) halves by cutting it between the eyes, through the middle of the body, and then through the tail. But if we cut across the abdomen from the left to the right side, one half would have a tail and the other half a head. Obviously these halves are not identical.

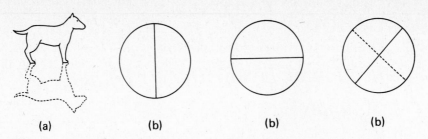

(a)   (b)   (b)   (b)

FIGURE 5-1   (a) Bilateral symmetry. Only a line drawn from front to back down the middle of the animal results in mirror images. (b) Radial symmetry. Any number of lines bisecting the animal result in mirror images.

The Radiata are characterized by the fact that they can be cut into identical halves in many ways, provided only that a straight line is drawn through the center of the animal (see Figure 5-1).

1. Examine the preserved organisms at your table. Name three animals with radial symmetry and three with bilateral symmetry.

Bilateral Symmetry   *Radial Symmetry*

_____   _____

_____   _____

_____   _____

Segmentation

Many animals have structures or body parts which are linearly repeated, like the cars of a railroad train. These parts are called segments. A clear example of this characteristic is the repetitive arrangement of the external body parts of the sandworm, *Nereis,* or the earthworm. In some animals the segmentation has been secondarily lost, or so modified that the repeated parts are no longer similar in appearance. It is sometimes possible to find clear segmentation in the embryo or in the adult internal anatomy. Look at the human skeleton. What evidence is there of segmentation?

2. List three segmented and three nonsegmented animals.

Segmented Animals   *Nonsegmented Animals*

_____   _____

_____   _____

_____   _____

Skeleton

Some animals, like the lobster or beetle, have exoskeletons (outer skeletons). There are no bones beneath their hard outer "shells."

3. What is an obvious function of such a skeleton?_____

4. Man and other vertebrates have internal skeletons. While bones like those of the skull are obviously of a protective nature, what additional function have bones in the vertebrate

body?_____

Explain how this works. _____

_____

5. Where are you likely to find the muscles of a lobster in relation to the exoskeleton?

_____   _____

How do you think they function in locomotion?_____

_____

_____

6. In addition to animals with exoskeletons and animals with endoskeletons, what other

category is there?_____ List three groups of animals in each of the first two categories and two groups fitting into the third one.

With Exoskeleton	*With Endoskeleton*	*Other*
_____	_____	_____
_____	_____	_____
_____	_____	_____

Paired Appendages

7. How many appendages are found in pairs in the following animals? (Don't count the pairs, count the appendages.)

     a. Man _____

     b. Crayfish _____

     c. Grasshopper _____

     d. Fish _____

     e. Bird _____

Using wooden probes if necessary, study the following internal characteristics on the dissected animals at the front table.

Digestive Tract

An animal can have (a) no digestive tract, (b) a sac-like digestive tract with only one opening, or (c) a tube-like digestive tract with a mouth and anus.

8. Name one animal with each type of digestive tract.

a. _____

b. _____

c. _____

Respiratory Tract

9. Name one kind of animal with each of the following.

a. Lungs _____

b. Gills _____

c. Spiracles (rows of little holes at the base of each leg or along the edges of the body

which let in air) _____

Nervous System

Some animals, such as jellyfish, have diffuse nervous tissue which is distributed throughout the body. Other animals may have no specific nervous tissue, responding to stimuli through the innate irritability of their protoplasm. Some animals, however, have very specific nervous tracts culminating in an anterior accumulation of nervous tissue. This is called **cephalization** (cephalic: having a head).

10. What are some external signs of cephalization? _____

_____

11. Name one kind of animal with each of the following.

a. No nervous system _____

b. Diffuse nervous system _____

c. Nervous system characterized by cephalization _____

The characteristics described above are more or less natural criteria which have become apparent to taxonomists after many years of observation of animal morphology. The physical characteristics of many animals were observed and recorded (accumulation of facts) and natural relationships were perceived (induction). These inductions, plus the search for order which is characteristic of science, led to the construction of a system of classification into which each kind of animal more or less naturally fits.

Once the criteria for these groups were established, it was then possible to look at newly discovered animals from the point of view of their specific categories. Furthermore, it became possible to look for new animals in habitats which could now be anticipated. For example, new worms would be searched for under the ground, new birds in the air, etc. We see here the progressive effect of induction, in that from the generalizations it produces new facts may be deduced.

■ THE PROBLEM: *To classify animals according to their physical characteristics.*

You will find the morphological categories which are used to differentiate animals listed in a key below. This is a device for progressively eliminating groups of animals from your list of "suspects," until only the category of the animal you wish to classify is left. You are to choose from alternative generalizations until the appropriate one is found. Remember that you are now working from the *general* (common characteristics of the group) to the *specific* (your particular animal). This is the process of deduction, which we have already defined.

Here are some examples of deduction: a mother predicting when her child will awaken from a nap on the basis of past experience: predicting the winner of a baseball game; searching for a new drug among molecules proven effective on related diseases; predicting the weather.

How to Use the Key

Select the appropriate alternative of the two (or three)—for example, **A** or **AA**. Now look at the right-hand column to find the next letter to go to. Continue following the letters until you get to a word. This is the phylum you are looking for. If there is no letter to the right of the phylum, you are finished. In the case of phyla Arthropoda and Chordata, you must go on to class, and a letter will follow the phylum.

12. Using the key, classify the following animals before you attempt to work on the unknowns.

Butterfly _____   Snake _____

Crayfish _____   Frog _____

Clam _____   Jellyfish _____

Starfish _____   Earthworm _____

Liver fluke _____   Fish _____

Sponge _____

You are expected to identify correctly all of the animals in the room as to phylum. In addition you are to learn the classes of phyla Arthropoda and Chordata. Your instructor will review the gradations of the binomial system of classification, from kingdom to genus and species.

Key for the Classification of the Major Groups of Animals*

Description	Phylum, Subphylum, or Class	Go to Letter
A Radially symmetrical		B
AA Bilaterally symmetrical; tubular digestive tract with mouth and anus usually present		C
AAA Body may be symmetrical or not; body porous, usually feels spongy	Phylum Porifera	
B Body soft and usually sac-like or umbrella-like; digestive cavity with only one opening	Phylum Cnidaria (synonym: Coelenterata)	
BB Hard body wall, spiny; body usually divided into five or multiples of five; digestive tract with two openings (mouth and anus)	Phylum Echinodermata	
C Never possess paired appendages; nonsegmented		D
CC Paired appendages frequently present; segmentation present but may be modified or not clearly evident		F
D Body flat or ribbon-like	Phylum Platyhelminthes (flatworms)	
DD Body not flat		E
E Body elongated, cylindrical, pointed at both ends	Superphylum Aschelminthes Phylum Nematoda (roundworms)	
EE Body soft, has flattened or branched muscular foot and mantle; many forms possess a shell	Phylum Mollusca	
F No skeleton; appendages, if present, are nonjointed; body soft, segments quite similar	Phylum Annelida (segmented worms)	
FF Skeleton present; usually possess paired, jointed appendages		G
G Skeleton internal; usually possess two pairs of jointed appendages	Phylum Chordata	K
GG Skeleton external; possess three or more pairs of jointed appendages	Phylum Arthropoda	H
H Segments similar; more than ten pairs of legs	Class Myriapoda	
HH Segments usually dissimilar		I
I Without antennae; head and thorax united; normally four pairs of walking legs, two pairs of legs modified for mouth parts; in some forms one of these pairs may be used as walking legs (usually has eight legs and a body divided into two parts)	Class Arachnida	
II With antennae		J

*Modified from Lawson et al., *Guide for Laboratory Studies/Biological Science*, 4th ed., Michigan State College Press, East Lansing, Mich., 1951.

	Description	Phylum, Subphylum, or Class	Go to Letter
J	One pair of antennae, three pairs of walking legs; usually possess wings; three body parts (head, thorax, abdomen)	Class Insecta	
JJ	Two pairs of antennae; five to seven pairs of walking legs; wings absent	Class Crustacea	
K	Without vertebral column; sac-like body with two openings (possess, during embryological development, pharyngeal gill slits, dorsal nerve cord, and a rod-like structure called a notochord)*	Subphylum Urochordata	
KK	Vertebral column; brain case; usually possess two pairs of appendages supported by internal skeleton	Subphylum Vertebrata	L
L	Gills present (sometimes present only in larval stage)		M
LL	Never possess functional gills; lungs present		N
M	Possess gills during entire life; body covered with scales	Class Osteichthyes	
MM	Lungs, rarely gills in adult; skin smooth, moist; legs present (eggs laid in water or moist places)	Class Amphibia	
N	Skin dry, scaly, without hair or feathers	Class Reptilia	
NN	Feathers	Class Aves	
NNN	Skin hairy; mammary glands present	Class Mammalia	

*Adults of this subphylum will be on display at the front table. The characteristics which make these animals chordates will not be visible, as they are primarily embryonic.

REPORT SHEET   Name _____

                              Instructor's name _____

Classification of Unknown Animals

At one end of your table you will find several numbered bottles containing specimens. Next to the appropriate number below, identify the phylum of the specimens found in each of the bottles. In the case of phyla Arthropoda and Chordata, indicate the class as well as the phylum.

After you finish examining each bottle, return it and obtain another. At no time should you have more than one bottle at your place.

This page will be collected at the end of the period and graded; *it will therefore be necessary for you to work alone on this problem.*

1. _____   11. _____

2. _____   12. _____

3. _____   13. _____

4. _____   14. _____

5. _____   15. _____

6. _____   16. _____

7. _____   17. _____

8. _____   18. _____

9. _____   19. _____

10. _____   20. _____

6 Deduction II

Classification of Living Things: Form and Structure 3: Kingdoms Monera, Protista, Fungi, and Plantae

In the past, all living things were placed into two kingdoms, Plantae and Animalia. But as our knowledge grew, it became apparent that two categories did not adequately represent all major groupings of organisms. For example, biologists knew of two species of one-celled organisms which looked identical. Both had a whip-like flagellum used for propulsion. One, *Peranema*, was considered an animal. The other, *Euglena*, was considered a plant because it had tiny capsules of chlorophyll (chloroplasts) inside its body and was able to carry on photosynthesis. But someone discovered a way of removing the chloroplasts. The chlorophyll-free *Euglena* swam around, not realizing that it had just undergone a monumental transformation. By removing its chloroplasts the scientist had converted *Euglena* into what appeared to be *Peranema*—a plant had been converted into an animal! Obviously, the two kingdom system was an oversimplification. A modern synthesis has resulted in the five kingdoms listed in Table 6-1.

TABLE 6-1   The Five Kingdoms

1. Multicelled animals   **Kingdom Animalia**
2. Some single- and many multicelled organisms containing chlorophyll
                                                          Kingdom Plantae
3. One-celled organisms including certain chlorophyll-bearing forms and those without chlorophyll   **Kingdom Protista**
4. Organisms whose main body mass consists of a tangled mass of threads, a mycelium, and which usually produce a fruiting body containing spores   **Kingdom Fungi**
5. Organisms consisting of single cells or chains of cells whose nuclear material is scattered throughout the cell, not aggregated into a discrete nucleus.   **Kingdom Monera**

Kingdom Fungi includes a variety of forms which are difficult to associate into one group. Some fungi are larger than a man's head; others produce tan, leathery, tennis-ball-sized and -shaped fruiting bodies; still others produce the classical mushroom. Yet some fungi, such as yeasts, consist of microscopic single cells which reproduce by budding. All lack chlorophyll. All absorb food on a molecular level and lack a mouth.

Kingdom Monera comprises the bacteria and the blue-green algae. These seemingly disparate groups do have much in common: both contain cytoplasm which lacks vacuoles. The cytoplasm appears more rigid than that of cells of the other kingdoms, and it is more resistant to damage by heating, drying, and chemicals. Both phyla have unusual molecules, mucopeptides, in their cell walls.

Notice that the criterion for inclusion in Kingdom Plantae is not simply the presence of chlorophyll. Certain one-celled organisms containing chlorophyll are considered plants, while others are considered protists. In some cases the differentiation is easy to understand. The Euglenophytes, for example, contain some species with chloroplasts, but none have cellulose cell walls, which are more or less universal among plants. They are thus considered protists. However, Phylum Chrysophyta, a group which often contains chlorophyll *and* has cellulose cell walls, is also placed amont the protists.

This exercise will consider these kingdoms plus some of the more common plant phyla and the classes and subclasses of the seed plants.

USE OF THE MICROSCOPE

The microscope is a tool used by scientists to extend their ability to see small objects. Since the microscopic world is not part of our usual visual frame of reference, it will be necessary for you to *know the characteristics of what you are looking for* before you can pick it out of the many shapes and sizes which meet your eyes when you look into the microscope. *Before you look into the microscope, make sure you have at least an idea of what you are looking for.*

Figure 6–1 is a labeled drawing of a microscope. Make notes below the drawing as your instructor explains how to use this instrument.

KINGDOM PLANTAE

Certain characteristics are used to distinguish the various divisions, classes, and subclasses from one another. In classifying a given plant, some or all of the following criteria are used.

(1) Does the plant have seeds or spores?
(2) Are the seeds in cones or in a part of the flower?
(3) Do the seeds have one cotyledon or two?
(4) Are the flower parts in threes or multiples thereof or in fours or fives?
(5) Does the plant have vascular bundles?
(6) If the stem has vascular bundles, are these scattered or in a ring?
(7) Are the leaf veins parallel or netted?
(8) Is green color (chlorophyll) present or absent?

Before attempting to classify plants, you must become familiar with the meaning and application of the terms used in this paragraph. On the front table you will find materials to illustrate the terms used.

Seeds and Spores

Only two classes of one of the nine phyla to be studied have seeds. All the rest reproduce themselves by means of spores. If you will stop to consider, you will realize that you have never seen the seed of moss, fern, or mushroom. These plants possess spores instead

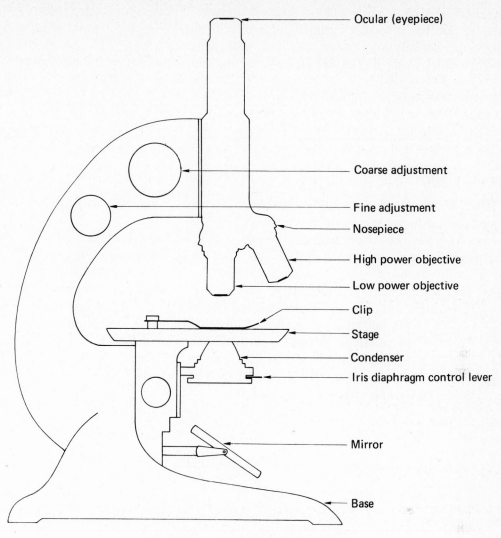

FIGURE 6-1 Compound microscope.

Notes

of seeds. A seed is a multicellular body with large amounts of stored food. A spore is unicellular and has a smaller quantity of available food. For this reason, seeds can remain dormant for relatively long periods of time, while spores must grow in a shorter time if they are to grow at all. Compare the size of the seeds and spores present on the laboratory table.

1. Describe the size difference. _____

_____

Cones and Flowers

If we consider only the classes with seeds (Gymnospermae and Angiospermae), we find that seeds can be borne in either cones or flowers.

Pine trees have cones composed of woody scales. At the base of each scale one can find a winged seed. When a ripe cone is shaken, the seeds easily sift out from among the scales. No tissue actually surrounds the seed. These plants are called gymnosperms or naked-seeded plants. The spruce, fir, hemlock, and other such plants are members of the class *Gymnospermae*. Examine the cones provided and observe the scales and seeds.

Next observe the pod of a pea or bean. Note that the seeds are surrounded by a pod and are not "naked." It is therefore necessary to "shell" the pea or bean pod to observe the seeds. These plants and many thousands similar to them are members of the class *Angiospermae* (seeds in a vessel). The vessel or pod is the enlarged base of the pistil of the flower, the other parts of the flower having withered and fallen off.

Cotyledons

Examine the soaked beans (or peas) and corn, and with your fingernail separate one seed of each into parts.

2. Which one separates easily into two equal halves? _____

3. Which one does not so separate? _____

Each separate part of the bean seed is called a cotyledon. Plants with seeds having two cotyledons are called dicotyledons, and those whose seeds do not separate and have only one cotyledon are the monocotyledons. The Monocotyledonae (monocots) and Dicotyledonae (dicots) are subclasses of class Angiospermae in the phylum Tracheophyta.

Flower Parts

Review by means of the charts and models in the laboratory your knowledge of the parts of a flower. Locate the sepals, petals, stamens, and pistil.

4. What is the most common number of parts of the snapdragon (one, two, three, four,

or five)? _____

5. What seems to be the lowest multiple of parts in the gladiolus? _____

The flower parts of monocotyledons usually occur in threes or multiples thereof, such as three sepals, three petals, and six stamens; those of dicotyledons occur in fours or fives.

Leaf Venation

The monocotyledons and the dicotyledons can usually be distinguished from one another by the type of veins that their leaves display. Veins which are found to lie parallel to one another from the base of the leaf to the tip are known as **parallel veins** and are characteristic of the monocotyledons. Veins which form various patterns resembling a net are called **net veins**. Such types are found in the dicotyledons.

Examine the corn leaves on your table, as well as the leaves of the geranium and other plants supplied.

6. Name the plants that belong in each of the following groups.

    a. Parallel-veined plants _____

    b. Net-veined plants _____

    c. Monocotyledons _____

    d. Dicotyledons _____

Vascular Bundles

Observe the small dots on the cut end of a corn stem. Take a portion of this stem about ½ inch long and crush it.

7. Are these dots parts of structures which extend lengthwise through the stem? _____

Observe a portion of a corn stem with a piece of the leaf attached.

8. Do the strings of tissue extend into the leaves? _____ Dissect the piece of corn stem if necessary to determine this.

These strings of tissue are called **vascular bundles**. When they occur in the leaves they are also called **veins**. A vascular bundle is a strand of tissue through which water, minerals, and food move. A bundle consists of **xylem** vessels, which carry water and minerals, and **phloem** sieve tubes, which carry food. In cross section, the xylem vessels are most easily distinguished. They appear to be comprised of cells whose walls are thick, angular in appearance, and stained red (by the dye used in making the slide). In longitudinal section these vessels appear as long, hollow tubes.

Vascular bundles are very important in the classification of plants. Their presence or absence must always be noted. Leaf-like structures without vascular bundles are not true leaves. Similarly, stem-like or root-like structures which do not have vascular bundles are not true stems or true roots. In other words, **all true roots, stems, and leaves possess vascular bundles**.

Examine under the microscope slides of corn, bean, fern, and mushroom stems.

9. List those with and those without vascular bundles.

    a. Vascular bundles present in stem   _____

    b. Vascular bundles absent from stem _____

Examine corn, geranium, fern, and moss for vascular bundles in their leaves or leaf-like structures. Hold the fern and moss up to a light source. Note that a single midrib is not evidence of vascular bundles (veins) in the leaf.

10. List those with and without vascular bundles.

    a. Vascular bundles present in leaf _____

    b. Vascular bundles absent from "leaf" _____

Vascular Bundles Scattered or in a Ring

You have already looked at the corn and bean stems and found the vascular bundles. Examine the slides of corn and bean under the microscope again and observe whether the vascular bundles are scattered all through the cross section or in a ring near the outer edge.

This difference, which should be apparent, is a useful one in classification. Monocotyledons have scattered bundles as well as parallel-veined leaves and flower parts in three or multiples of three. Dicotyledons have vascular bundles in a ring, net-veined leaves, and flower parts in fours or fives.

11. Which plant has scattered vascular bundles?_____

12. Which plant has vascular bundles in a ring?_____

Chlorophyll

Plants are usually thought of as green organisms, but this is true only in part. The common green plants such as grass, trees, mosses, ferns, and green algae do contain a green substance called chlorophyll, but sometimes the green may be masked by another pigment. Further, the green color in green molds is not chlorophyll. In classifying plants you will have to determine whether chlorophyll is present or absent.

■ THE PROBLEM: *To classify members of Kingdoms Plantae, Protista, Fungi, and Monera according to their physical characteristics.*

13. When you use the key to determine the classification of an organism, are you performing induction or deduction? _____

At each table are several bottles and boxes of dried or otherwise preserved organisms. Identify them according to kingdom or phylum. Tracheophytes should be keyed down to class and Angiosperms to subclass. As you become more proficient in keying out organisms you will need to use the key less and less until, finally, you will be able to discard it.

Key for Classification of Kingdoms Plantae, Fungi, Protista, and Monera

Description	Kingdom, Phylum, Class, or Subclass	Go to Letter
A Single cell		**B**
AA Many cells arranged in strands, tissues, or tissue-like masses		**E**
B Cell contains a nucleus or has glassy, symmetrical shape without apparent cellular contents	Kingdom Protista	**D**
BB Cell without discrete nucleus, nuclear material scattered throughout cell. Cells appear as minute rods, dots, corkscrews or bluish green cells or chains of cells	Kingdom Monera	**C**
C Tiny; visible under oil immersion lens of microscope as dots, rods, or corkscrews	Phylum Schizophyta (Bacteria)	
CC Individual blue-green cells, colonies, or chains of cells. Not identifiable with classroom equipment.	Phylum Cyanophyta (Blue-green algae)	
D Cell lacks pigment in discrete capsules (chloroplasts).* Nucleus may be stained (usually purple) and cytoplasm may be counterstained, but lacks green pigment. May have whip-like flagellum, hair-like cilia, or temporary lobe-like locomotor appendages (pseudopodia)	Phylum Protozoa	
DD Cell appears to lack any contents; glass-like wall with symmetrical snowflake, crescent, or petri-dish shape. (Preparations of these cells have been bleached so that only the glassy cell walls are visible. When alive, cells contained a nucleus and golden brown or yellow-green pigments including some chlorophyll pigments.)	Phylum Chrysophyta (Diatoms and related algae)	
E Chlorophyll present in discrete capsules: color may be yellow-brown, red, or green	Kingdom Plantae	**F**
EE Chlorophyll absent. Usually composed of filaments of cells visible under low magnification. Fruiting body visible to naked eye as a dot or may appear mushroom-shaped. Reproduce by spores.	Kingdom Fungi (Mushrooms, molds, yeasts, etc.)	
F With true roots, stems, and leaves (containing vascular bundles).	Phylum Tracheophyta	**J**
FF Lacks true roots, stems, and leaves. Has small leaf-like structures and grows close to ground, or may have large bush form. Examination of leaves reveals absence of venation, although a midrib may be present.		**G**
G Green; with small leaf-like structures; rarely grows higher than 2.5 cm (1 in.). Often forms green carpet.	Phylum Bryophyta (Mosses)	
GG Green, yellow-brown, or red. May form chains of cells, masses of green filaments, bushy plants. (May also exist as individual cells; to avoid confusion such cells are not included in specimens you will see.)		**H**

*Some protists have chloroplasts, but none of these will be found in your classroom materials.

Description	Kingdom, Phylum, Class, or Subclass	Go to Letter
H Green; cells in long strands, spherical colonies, or single individuals. Spiral or capsule-like chloroplasts. Usually microscopic or forming tangled masses of threads or sheets of cells in salt or fresh water.	Phylum Chlorophyta (Green Algae)	
HH Yellow-brown or red. Can form large rubbery or filmy plants. Chlorophyll present but not usually visible in chloroplasts.		I
I Yellow-brown; may have root-like and/or leaf-like structures, but close examination reveals no veins (though a supporting midrib may be present).	Phylum Phaeophyta (Yellow-brown algae or seaweeds)	
II Pinkish to bright red. May have root-like and/or leaf-like structures but lacks true venation. Can be large, filmy, sheet-like, or bushy	Phylum Rhodophyta (Red Algae or Red Seaweeds)	
J Reproduce by seeds which arise in cones or flowers	Subphylum Spermopsida	K
JJ Reproduce by spores; no seeds, cones, or flowers. May have brown, ring-like, spore-producing structures on underside of leaves.	Subphylum Pteropsida (Ferns)	
K Seeds produced in cones; leaves usually needle-like.	Class Gymnospermae (Conifers—pine trees and relatives)	
KK Seeds developed from flowers; leaves usually broad (not needle-like).	Class Angiospermae (Flowering plants)	L
L Seeds with one cotyledon; leaves usually parallel-veined; flower parts in threes; vascular bundles in stem scattered.	Subclass Monocotyledonae	
LL Seeds with two cotyledons; leaves usually net-veined; flower parts in fours or fives; vascular bundles of stem in a circle.	Subclass Dicotyledonae	

After practicing with the key for a while, consult the flow chart (Figure 6-2) for an overview which might help you organize the taxonomy in your mind.

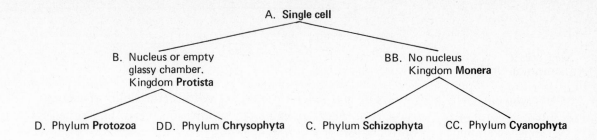

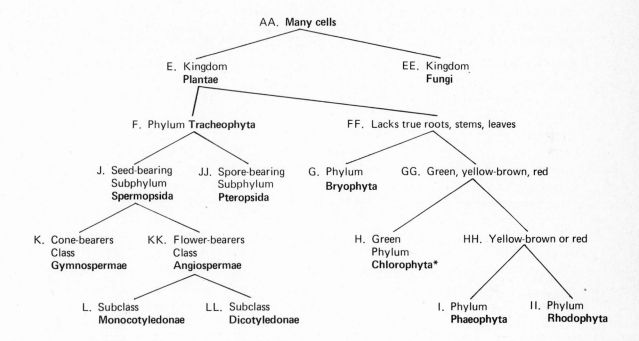

*May be one-celled.

FIGURE 6-2   Flow chart of the taxonomy of kingdoms Plantae, Fungi, Protista, and Monera.

REPORT SHEET   Name _____

                          Instructor's name _____

Classification of Unknown Plants

At one end of your table you will find several numbered bottles containing specimens. Classify them according to division and, where possible, to subdivision, class, and subclass. After you finish examining one bottle, return it and obtain another.

1. _____

2. _____

3. _____

4. _____

5. _____

6. _____

7. _____

8. _____

9. _____

10. _____

11. _____

12. _____

13. _____

14. _____

15. _____

7 Teleology

Cells and Cellular Differentiation

If we are to be erect, we need structure to support the body. If we are to think, we need structure to create and transmit thoughts. If we are to move, we need structure to initiate and sustain movement. An understanding of the evolution and development of the cells, tissues, and organs of the body begins with an appreciation of the nature of the problem of cellular differentiation. Your body began as one cell, the fertilized ovum. What tremendous potential there must have been in that cell to result in the mass of bone, muscle, nervous, and connective tissue that represent an adult human being!

At the present stage in the development of biological theory, we vaguely understand that various groups of cells are capable of secreting substances which initiate the differentiation of nearby cells. These cells in turn begin to secrete substances which nullify the effect of the original secretions and begin to stimulate still further changes in the morphology of other nearby cells. Each differentiating mass of cells is controlled by a separate center or focus of secretion called an **organizer**; each mass of cells has its own feedback effect on the organizer, which in turn causes the organizer to alter its original function.

One gets an inkling of the complexity of this amazing phenomenon from following the embryonic development of the eye. As the optic vesicle extends from the brain and grows toward the surface of the head, it begins to invaginate to form a cup. The inner layer of this cup will be the retina, the sensory area of the eye. At the same time that the cup is developing, certain cells making up the skin on the outside of the head of the embryo begin to form a lens. Finally, the optic cup grows to a point *just under the lens*, and the two structures fuse to become the eye.

One is tempted to ask the questions Why? and How? when faced with this phenomenon. The answer to "how" can be approached, albeit with incomplete explanations, by the introduction of concepts like the "organizer"; "why" is a more difficult problem. Why are some cells elongated and springy, while others secrete a solid matrix around themselves to form a hard tissue? Why are nerve cells long and thin and often insulated so that they do not lose the nerve impulse? Why do red blood cells have hemoglobin, without which we would be unable to transport enough oxygen through the body to sustain our life processes?

One is tempted to answer (1) "If our cells did not develop such variations, we would be unable to function" or (2) "If we didn't have cells to perform their respective functions, we would die." But are such conclusions of real value in furthering our knowledge of the reasons for cellular differentiation? Before reading on, answer this question.

1. What is unproductive about the two preceding answers to the question "Why do cells

differentiate?" _____

_____

_____

_____

Your response should have been that such answers are **teleological**, which means that to answer in the preceding manner is to confuse cause with effect. Science is based on the tenet that every effect has a cause. A scientist may spend many years studying a phenomenon without more than a clue as to its cause because he believes implicitly that it must have one. Often, however, we are led astray by answers which seem correct but which really do not fit into this cause-and-effect relationship. Sometimes a question requiring a **cause** for an answer is answered by an **effect**. For example, if we ask "Why do cells differentiate?" we are asking for a cause for the effect called cellular differentiation. To reply "because otherwise we would die" is to describe the *effect* of a lack of cellular differentiation. We have responded with an effect when we have been asked for a cause. A more valid answer would be "because they are subjected to the secretions of organizers."

When we combine the question with the answers,

(1) Cells differentiate because if they did not, we would die (teleological).
(2) Cells differentiate because they are subjected to the secretions of organizers (which in some way cause them to change their function, morphology, etc.).

the teleological answer can be recognized as an inadequate cause for the effect being studied.

The end product of a teleological orientation is the belief that the effect is implicit in the cause, or that each cause is predetermined to have a specific effect. This implies that all nature is goal-directed or aimed toward a particular purpose; for example, a seed becomes a plant, not because of a series of interacting processes, but because it is predestined to become a plant.

> Evidently, teleology "explains" the end state by simply assuming it given at the beginning. And in thereby putting the future into the past, the effect before the cause, it negates time. Teleology (often called "finalism") clearly leans toward the supernatural.*

The supernatural aspect of teleology made it useful to seventeenth century religious philosophers who sought to prove the presence of God by showing how miraculously the parts of the world fitted together. (Since the world worked so wonderfully, was this not proof it was controlled by a supernatural being?) Nehemiah Grew, in 1701, wrote

> The manifold variety of ears illustrate how the creator has adapted each animal's structure to the position it must fill. Since the owl sits on branches and listens for things below, its ear projection is above the auditory canal; while the fox which scouts for roosting prey has ears which are directed upward. The rabbit, always fearing pursuit, has ear passages turned to the rear; and so the horse who must listen for the driver behind.†

Grew was suggesting that the *cause* for the arrangement of ears was the use they are put to. But he ignored the fact that their use (the horse being able to hear the commands of the buggy driver) was really the *effect* of having such ears in the first place. This teleological reasoning reached its peak in a further discourse of Grew's on the horse.

> Because it must have huge lungs, the horse must have a broad chest, well-bowed ribs and wide nostrils. Since it is pestered with flies it must have a bushy tail to flick them off, while the thick-skinned donkey, little bothered by flies, neither needs nor has such a tail.‡

*Paul B. Weisz, *The Science of Biology*, McGraw-Hill, New York, 1959, p. 592.

†Nehemiah Grew, *Cosmologia Sacra: or a Discourse of the Universe as It Is the Creature and Kingdom of God* (London, 1701), p. 23.

‡Nehemiah Grew, *Musaeum Regalis Societatis. Or a Catalogue and Description of the Natural and Artificial Rarities Belonging to the Royal Society and Preserved at Gresham College* (London, 1681), preface.

To be satisfied with Grew's explanations the observer must believe that everything came into existence at the same time—that there was no evolution. The horse, his ears, and his driver are all part of one system where there is no necessity of one adapting to the other. Cause and effect are blurred—even unnecessary. The Creator initiated all parts of the system together. Westfall explains the flaws in Grew's reasoning.

> Grew's procedure involved a gross misuse of the scientific method. While seemingly he was basing his argument on observed empirical facts, he was neither inducing a conclusion from a body of observations nor employing systematic observation to check a theory. He was selecting facts to support a conclusion that he already held . . . he did not consider the unpleasant aspects of nature, the monstrous births, the diseases. . . . Those facts that he used were forced into an arbitrary pattern without the author's proving that they really fitted. Any conclusion was possible and facts were available to support any preconception when he could decide that the Creator turned the horse's ears toward the rear in order for it to hear the driver's commands.*

It should be apparent that teleological thinking is not productive, in terms of clearly establishing cause-and-effect relationships. Scientists believe that phenomena are sequential, each event in some way affecting the next. A seed does not represent a grown plant. Instead, the seed represents the vehicle and substance of a series of interacting reactions *resulting* in the grown plant. By examining processes in the light of their sequential development, the scientist is stimulated to try to understand their interaction. A teleological approach represses curiosity, for it suggests that the cause and the effect are both part of a preordained plan.

2. Put a T after each of the statements below that is a teleological conclusion.

   a. A democratic America means a strong America. _____

   b. The reason eggs have yolk is that they may provide food for the developing embryo. _____

   c. The plant grows upward because it needs the sunlight. _____

   d. My automobile starts up easily in the morning because otherwise I might not get to work on time. _____

   e. A stitch in time saves nine. _____

   f. Muscle cells are long and flexible and thus they can contract readily to cause movement. _____

   g. I am going to sleep because I am tired. _____

   h. The ancestors of birds evolved wings so that they could fly. _____

   i. Nerve cells are wire-like because they have to carry nerve impulses from one part of the body to another. _____

   j. The baby cried because it was hurt when it fell. _____

*Richard S. Westfall, *Science and Religion in Seventeenth Century England*, Yale University Press, New Haven, 1958, p. 62.

k. The plane's engines stopped because it ran out of fuel.   _____

l. The purpose of the heart is to pump blood.   _____

■ PROBLEM 1: *What is the morphology of the cell?*

The discovery of the microscope in the late sixteenth century made it possible for men to examine a whole new world of microscopic organisms and finally to understand that *all living things are made up of cells.* This latter statement was the culmination of a hundred years of study. What seems like a simple rule represents the synthesis of thousands of individual observations into a generalization of high magnitude, a **theory.** Subsequent investigations over more hundreds of years have shown the cell theory to be valid (with the usual exceptions to any scientific rule—remember, in science, there are no absolute truths), and it is now ready to achieve the status of a scientific **law.**

Cells differ drastically in appearance. Some are round, some are square, some have no constant shape. Cells may be found containing pigments of any color—or may be colorless and transparent.

Do cells have any characteristics in common, or is each of the thousands of different types of cells completely different from the others?

From the trays on the front table obtain slides of

> *Amoeba proteus* (a protist)
> *Elodea* leaf cells (an aquatic plant)
> *Spirogyra* (a green alga)
> Onion root tip

In addition, prepare a wet mount of human epithelial cells as follows.

Obtain a clean, blank slide, a wooden applicator stick, and a cover slip. Gently scrape the inside of your cheek with the applicator stick and *roll* the white material obtained onto the surface of the slide. Add a drop of iodine solution from the dropping bottle at your table. Gently place a cover slip on the preparation.

Examine each slide under low and high powers, devoting about 5 minutes to the study of each cell type.

Fill in the chart with descriptions of the aspects of cells listed in the top row. After completing the chart, use this information in the discussion led by your instructor, who will help you to learn the names of the parts of cells and to pinpoint the common aspects of cell morphology.

3. Do all cells have the same shape? _____ Proof. _____

_____

_____

4. Do all cells have the same internal structures? _____ Proof. _____

_____

Characteristics of Cells

	Shape	Is the outer boundary thick or thin? (Use high power.)	Contents of Cell
Amoeba proteus (animal cells)			
Human epithelium (animal cells)			
Elodea leaf (plant cells)			
Spirogyra (plant cells)			
Onion root tip (plant cells)			

5. Are there any characteristics which are common to animal cells but not to plant cells? _____

Characteristics of Plant Cells	*Characteristics of Animal Cells*

Note: Complete these two lists of characteristics after your class discussion of cells.

Draw a generalized cell in the space below and label all parts. Take notes below your drawing. Make sure you can identify the appropriate parts in the cells you have already observed and in any cells you see in subsequent sections of this exercise.

Drawing of a Generalized Cell

THE MODERN CONCEPT OF THE CELL

You have used an ordinary light microscope, an instrument unchanged in its essential powers of magnification since the seventeenth century. If your observations were made carefully, your drawing contains about as much information about cellular organelles as was known at the beginning of this century. But the invention of the electron microscope provided a breakthrough in visualizing the contents of cells. The observer is no longer dependent on the passage of light through thick glass lenses (as lenses get thicker, the potential for distortion becomes greater until, at between 1000X and 2000X, the practical limits of the light microscope are reached). The electron microscope provides almost unlimited opportunities to magnify small objects—reaching over 100,000X and making possible the examination of objects even at the molecular level (see page 312). It has thus become possible to visualize heretofore invisible organelles inside the cell—those structures that actually perform many of the processes of life.

Figures 7-1, 7-2, and 7-3 are electron micrographs of parts of cells magnified more than 10,000 times. Examine the photographs carefully, identify the various organelles (A-I) described below, and label each photograph appropriately. Then draw organelles A-I into your drawing of a cell on the preceding page.

A. **Nuclear membrane:** Outer membrane of nucleus. Consists of two layers contiguous at one or several points with endoplasmic reticulum. Controls passage of molecules into and out of the nucleus. Nuclear pores, spaces in the membrane, allow passage of large molecules into and out of the nucleus.

B. **Endoplasmic reticulum:** A network of double membranes forming a complex system of tubules and sacs, often studded with dot-like ribosomes (rough endoplasmic reticulum). Enzymes are manufactured here under direction of RNA from nucleus.

C. **Golgi apparatus:** Folded membranes containing substances to be processed and then secreted or excreted from cell. Often looks like a small, more or less circular, stack of membranes or layers of elongated folded membranes.

D. **Nucleus:** Round or oval body in cell. May or may not contain a nucleolus. Contains, among many other molecules, DNA and proteins, which may be distributed as thin, invisible threads contracted into distinct, visible chromosomes. The nucleus controls the metabolism of the cell by sending instructions for specific enzymes to be manufactured by ribosomes on the endoplasmic reticulum.

E. **Nucleolus:** A more or less spherical organelle inside the nucleus which contains RNA and proteins. Site of ribosomal RNA synthesis.

F. **Mitochondrion:** An oval organelle consisting of numerous shelf-like horizontal layers (cristae) responsible for producing energy to power cellular processes. The mitochondrion releases the energy from sugar molecules and stores it in ATP molecules.

G. **Lysosome:** Variably shaped sacs, each about the size of a mitochondrion, containing protein-digesting enzymes. Usually destroy damaged organelles inside the cell. In dying cells, the lysosomal membrane breaks down, and its protein-digesting enzymes destroy the cell from inside.

H. **Chromosome:** Thread-like strand of DNA and protein containing the instructions which are picked up by RNA molecules and carried to the ribosomes of the endoplasmic reticulum. Chromosomes usually appear in groups, as dark, elongate, irregular masses in the nucleus—but are visible only when cells are in the process of dividing.

I. **Ribosomes:** Various dot-like structures, primarily found lined up on membranes of endoplasmic reticulum. Composed of protein and RNA, these organelles are responsible for synthesis of proteins and, eventually, of the enzymes which control life functions at the molecular level.

Cell theory states that all living things are composed of cells and that cells are the basic unit of life.

6. If cell theory, as originally conceived, included cells with only the organelles present in your first drawing, does that mean that the addition of organelles A–I makes the original

theory wrong? _____ Explain. _____

_____

It should be apparent to you that theories are constructed with the understanding that they may become inadequate as more knowledge is accumulated. In fact, one purpose of a theory is to point out new directions for research, so that the theory may be further clarified. Science has the flexibility it needs because we recognize that even the most "sacred" theories may be modified as future knowledge reveals their flaws or limitations. There is no belief in "absolute truth" in science.

Your instructor will initiate a discussion of cells to help you fill in any blank spaces on the preceding pages.

Instructions for Labelling Photomicrographs

Write in, on the electron micrographs, all of the italicized terms in the caption for each photograph, filling in all of the spaces. Use the descriptions on page 75 to help you identify each organelle.

FIGURE 7-1 (opposite) Salivary gland secretory cell, magnified 7000X, showing organelles responsible for protein synthesis.

mRNA leaves (1) *nucleus* bringing instructions to ribosomes (barely visible as tiny dots) on (2) *rough endoplasmic reticulum*. Proteins are synthesized on the ribosomes and transported to a (3) *Golgi apparatus* for further processing. They are then stored, in the case of secretory cells such as this one, in (4) *secretory granules* (dark gray spheres). Energy is required for protein synthesis; thus there are many (5) *mitochondria* visible. The largest dark mass in the nucleus is the (6) *nucleolus*, where RNA is stored.

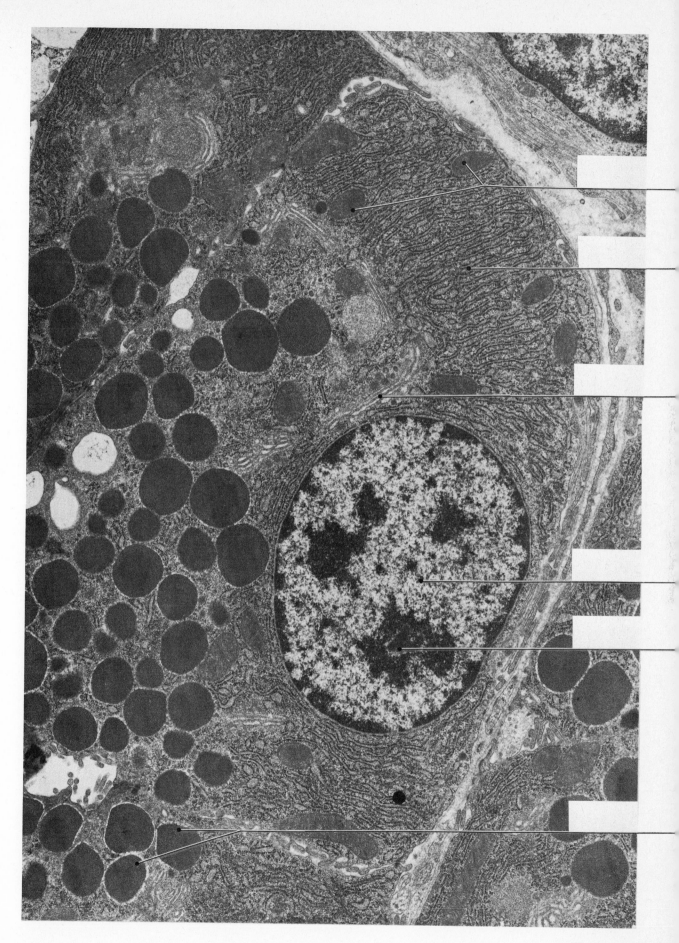

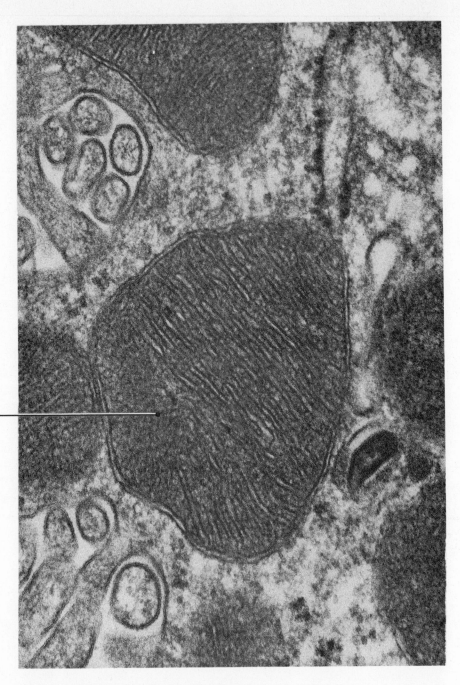

FIGURE 7-2 Cellular organelles.
(a) Notice the shelf-like structures on this highly magnified organelle responsible for producing energy in the cell. Label this organelle.

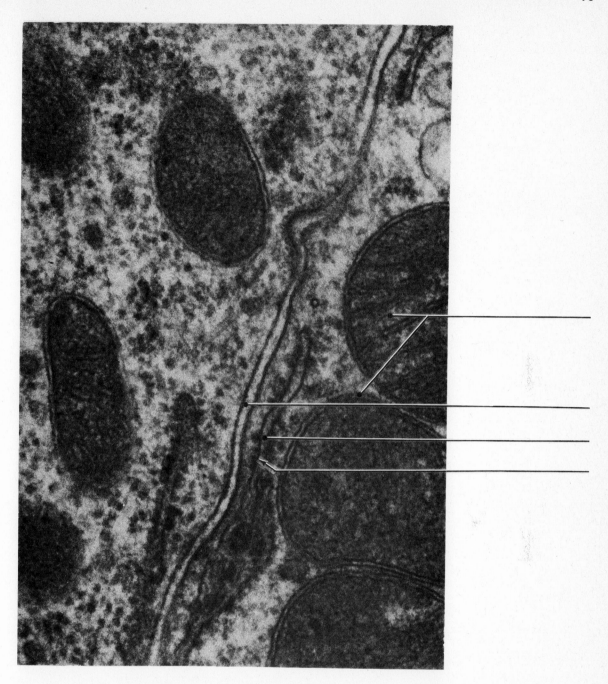

(b) The double-layered (1) *cell membrane* cuts diagonally across the center of the photograph. To its right is a thin area of (2) *rough endoplasmic reticulum*, showing dot-like (3) ribosomes. Label also (4) the large oval organelles at the right.

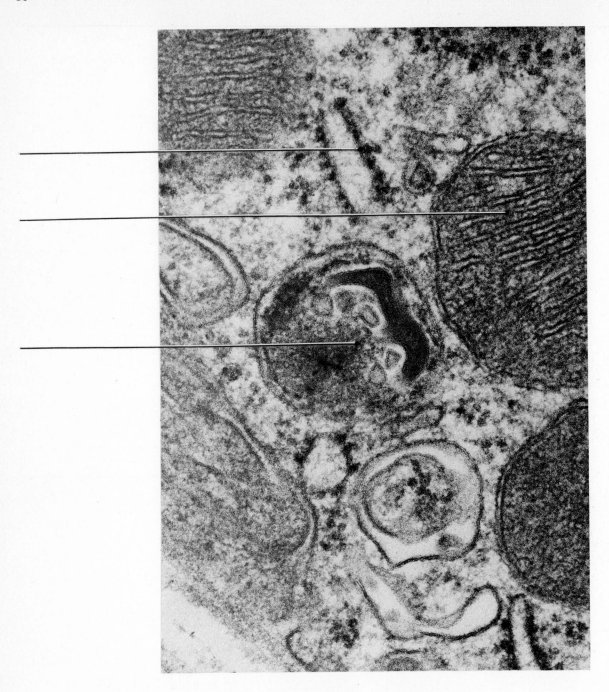

FIGURE 7-2   Cellular organelles *(continued)*

(c) Visible in the upper portion of the photograph are two rows of (1) *ribosomes* on a small piece of endoplasmic reticulum, beneath which is a thin-walled (2) *lysosome* with a dark mass inside it. Label also (3) the organelle at the right.

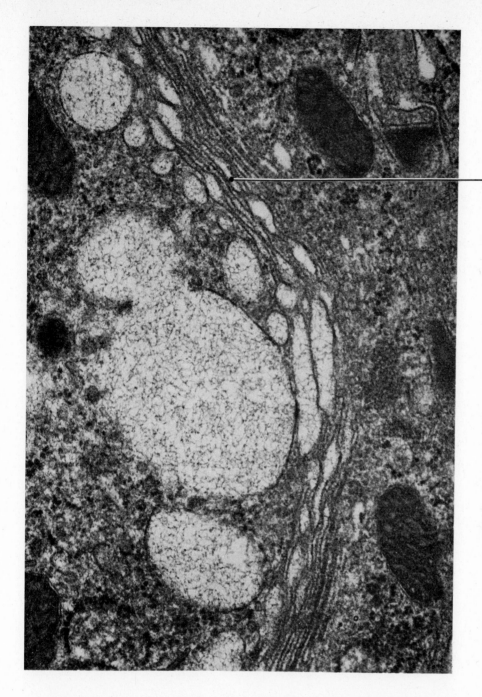

(d) The stacked layers of double membranes are part of an elongate *Golgi apparatus*. Label it. Note that in the elongate form it appears similar to endoplasmic reticulum, but it lacks ribosomes and its membranes are thinner (this photomicrograph is highly enlarged).

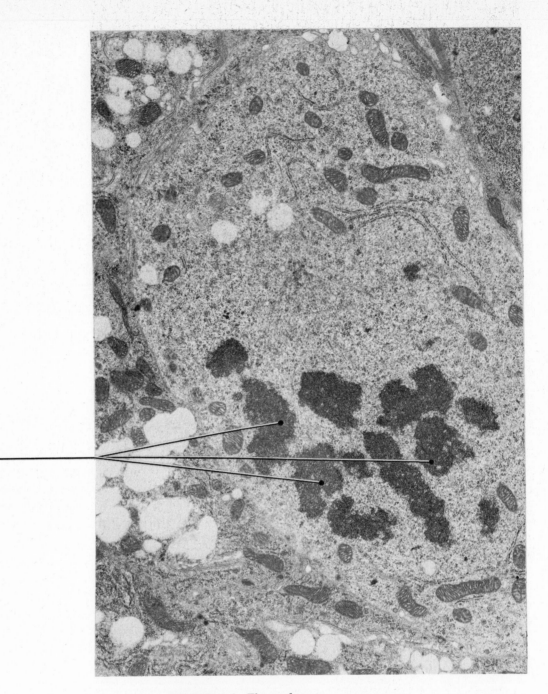

FIGURE 7–3 The nucleus.

(a) Label the large, dark, irregular *chromosomes* in the nucleus of this cell. Compare the sizes of the chromosomes and the abundant mitochondria.

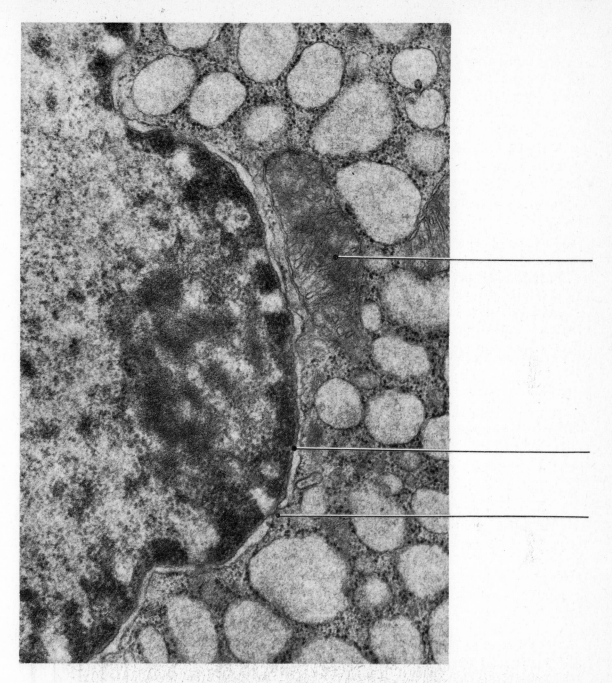

(b) This very highly magnified view shows part of the nucleus and its double (1) *nuclear membrane*. Find the (2) *nuclear pores* in the membrane, which permits passage of large molecules. Notice the circular rough endoplasmic reticulum in the cytoplasm, recognizable by many dark dotlike ribosomes; endoplasmic reticulum is not always elongate. A large (3) *mitochondrion* lies alongside the edge of the nucleus.

■ PROBLEM 2: *To follow the transition from single cells to multicelled organisms.*

The human body is composed of millions of cells, all of which were derived from one cell, the fertilized egg. How does this process occur? Historically, two major alternatives were presented. One theory, called the **preformationist doctrine**, suggested that the sperm or egg contained a tiny, fully formed human (called a **homunculus**), which simply grew to the size of the newborn infant inside the mother. Many famous early microscopists espoused this idea. In fact, soon after the discovery of the microscope, reports came pouring in describing observations of tiny humans seen in sperms and eggs, with corroboration in the form of reports of tiny roosters in rooster sperm, horses in horse sperm, etc.

Dalenpatius, a famous early scientist, wrote this description of human spermatozoa in 1699.

> They move with wonderful rapidity and by strokes of their tails produce little waves in the substance in which they swim. But who would believe that in these a human body was hidden? Yet we have seen such with our own eyes. For while we were observing them attentively, a large one threw off its surrounding membrane and appeared naked, showing distinctly two legs, thighs, breasts and arms . . . it was a delightful and incredible sight.*

7. Do you think that the preformationists were men of little integrity and were trying

to mislead the public? _____What is an alternative to this explanation? In

other words, how can you account for these seventeenth century "observations"?_____

_____

_____

The other theory which attempts to explain the development of the organisms from the egg is called **epigenesis**. This theory, as currently interpreted, suggests that the fertilized egg contains a set of instructions (a "blueprint") for the development of the organism and that these directions, passed down from parent cell to daughter cell, cause the production of substances which act on the cells, modifying them to their characteristic shapes and roles.

8. If you were living in the seventeenth century and had the foregoing two alternatives presented to you (with the "corroborating evidence" of the performationists), which would

you tend to believe? _____

You will be given the opportunity to use a modern microscope to check the validity of the preformationist doctrine.

Obtain a slide from the tray labeled *Starfish Development, Composite.* Examine it under the low power of your microscope. On the slide are all the developmental stages of the starfish from the unfertilized egg to the fully developed larva.

9. If the preformationist doctrine is correct, what should you be able to see in the

fertilized starfish egg?_____

*Dalenpatius as quoted by Vallisneri, from a paper by E.G. Butler in Knobloch, I.W., *Readings in Biologic Science*, Appleton-Century Crofts, New York, 1948 after Lewis and Stohr, *Text-Book of Histology*, Blakiston, 1913.

10. Find each of the stages mentioned below and draw a sketch of each.

 a. The unfertilized egg (one cell with a jelly-like "shell" around it and a prominent nucleus containing a dot-like nucleolus).

 b. The fertilized egg. Look for the disappearance of the nucleolus and the formation of a **fertilization membrane** between the jelly coat and the surface of the egg.

 c. The two-celled stage.

 d. The four-, eight-, and sixteen-celled stages.

 e. The 32-celled stage or **morula**, a solid ball of cells.

 f. The **blastula**, a hollow ball of cells.

g. The **gastrula**, a hollow ball of cells with a depression in it. This stage shows the beginning of differentiation in the starfish, because you can find two clearly defined layers from which the organs of the adult will be derived. One layer, the **ectoderm**, consists of the cells of the outer wall of the gastrula. From this layer, the skin, the nervous system, and certain other structures of the starfish will be developed. The other layer, the **entoderm**, is comprised of the cells lining the depression. This layer will form a third layer, called the **mesoderm**. The entoderm and mesoderm will form the internal organs and tissues of the starfish. In your drawing, label the ectoderm and entoderm. The mesoderm is not yet formed. It will appear between the ectoderm and entoderm.

h. The larva. At this point the three germ layers have developed into various tissues and clearly defined organs. After several days of free-swimming existence, the larva will undergo several rapid changes in appearance, settle on the bottom, and develop the characteristic features of an adult starfish.

11. Re-examine the fertilized egg. Can you see anything inside it which gives you a clue to its potential to become an organism of millions of cells, many of which will have no

resemblance to it? _____

12. Re-examine your drawings of the development of the starfish. Can you find any

definite clues to the mechanism which controls the changes in appearance of the cells? _____

It should now be apparent why the study of development is one of the most challenging and exciting aspects of biology and why there have been so many mystical and erroneous interpretations of this process.

Tissues are masses of similar cells. Your instructor will show 2 X 2 slides of various tissues. You are to sketch each tissue and learn to recognize the different types. It will be helpful if you associate the functions of the cells with their physical characteristics, provided you understand that the cells developed these characteristics not because the body needed them but as the effect of certain physiological and evolutionary processes.

When you have finished drawing the tissues, label the photomicrographs in Figure 7–4 (pages 90–91).

Name of Cell	Sketch (Draw each cell one inch in diameter)	Function and Distinguishing Characteristics
1. Adipose		
2. Smooth muscle		
3. Skeletal muscle		
4. Cardiac muscle		
5. Cartilage		
6. Bone		

Name of Cell	Sketch (Draw each cell one inch in diameter)	Function and Distinguishing Characteristics
7. White fibrous connective		
8. Nerve		
9. White blood cells		
10. Red blood cells		
11. Epithelial		

Some of the questions on your next practicum will be concerned with the identification of tissue types on microscope slides. Spend the remaining time of this period examining the microscope slides of each tissue type found in trays on the front table. If you do not have enough time to finish, come in on your own. *Make sure you have studied each tissue type carefully*. To help you identify the tissues, the photomicrographs in Figure 7–4 and the following descriptions are provided. The descriptions of the tissues are not in the order of the photographs, so you must study the text and pictures. When you find the description which fits the photograph, write the tissue name in the space below the micrograph.

Blood (high power): Mature red blood cells (erythrocytes) lack nuclei. Light-colored circles in a number of cells are depressions in each surface of the disc-like cell. Although all red blood cells are biconcave, the light-colored circles are apparent only when the cell is reflecting light at a certain angle.

White blood cells (leucocytes) are usually larger and much less abundant than red blood cells. This view shows one of the most common white blood cells, a neutrophil, with a trilobed nucleus. The plasma, or liquid portion of the blood, is invisible; it is a clear, watery fluid which has dried up in this preparation.

Bone (low and high powers): This tissue, like the other types of connective tissues, has a ground substance or *matrix* which fills in all spaces between cells. The cells are distributed in circles around a central *Haversian canal*, which contains an arteriole and a veniole. Food, wastes, and oxygen diffuse out of the blood vessels and cells and travel through tiny *canaliculi* (hair-like canals), since the matrix itself is relatively impermeable. Each group of bone cells (osteocytes) forming concentric rings around a Haversian canal is called a *Haversian system*. The low power photomicrograph on the left shows at least three Haversian systems. In this preparation the bone has been ground to a thin sliver and the cells have been destroyed. Each black hole is a *lacuna*, a space which formerly contained an *osteocyte*.

White fibrous connective tissue (high power): Numerous fibers criss-cross the field. The black dots are nuclei of oval cells, each of which has a fiber extending from each end. The gray background is the matrix.

Cartilage (low and high power): Cartilage cells (chondrocytes) appear in spaces (lacunae) in the gray matrix. A black nucleus can be seen in each *chondrocyte*. Cartilage is a type of connective tissue.

Skeletal, voluntary, or **striated muscle** (high power): Long, fiber-like, cylindrical cells marked with cross striations and numerous oval nuclei lying near the surface of the cell. In this preparation the striations are lightly stained, so look carefully for them.

Cardiac muscle (high and low power): This type of muscle tissue is found only in the heart. It consists of long, fiber-like cells with striations, similar to skeletal muscle, but with the following exceptions: It has fewer nuclei and they are located deep within the cell, not as near the surface as in skeletal muscle; the cells usually branch and do not lie as parallel as voluntary muscle cells; cells are separated from each other not by the usual cross membranes but by *intercalated discs*. These are tightly compressed and folded membranes which nerve impulses traverse freely, permitting the rapid contraction of heart muscle. An intercalated disc appears as a dark transverse line barely visible on the lowest fiber in the high power micrograph. Normally cardiac tissue is less distinctly striated than voluntary muscle.

Smooth muscle (low and high power): The left (low power) photograph shows a typical sheet of many smooth muscle cells, such as could be found in the small intestines or stomach walls. The high power view shows individual spindle-shaped cells teased apart from one another.

Adipose (fat) tissue (low and high power): Cells consist of one huge vacuole containing stored fat globules which push the cytoplasm and nucleus to the periphery of the cell

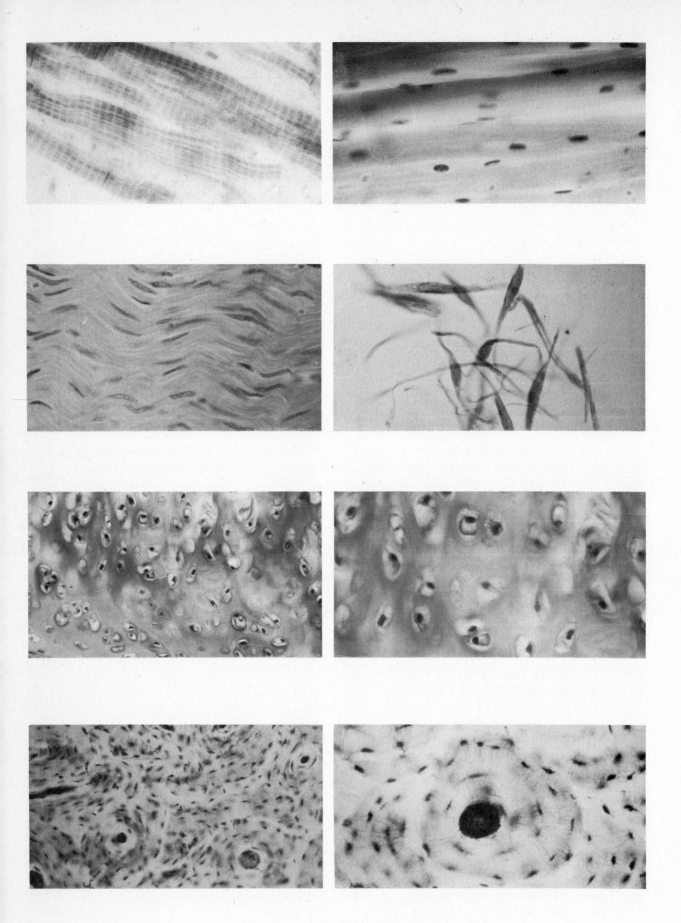

FIGURE 7-4 Human tissue types.

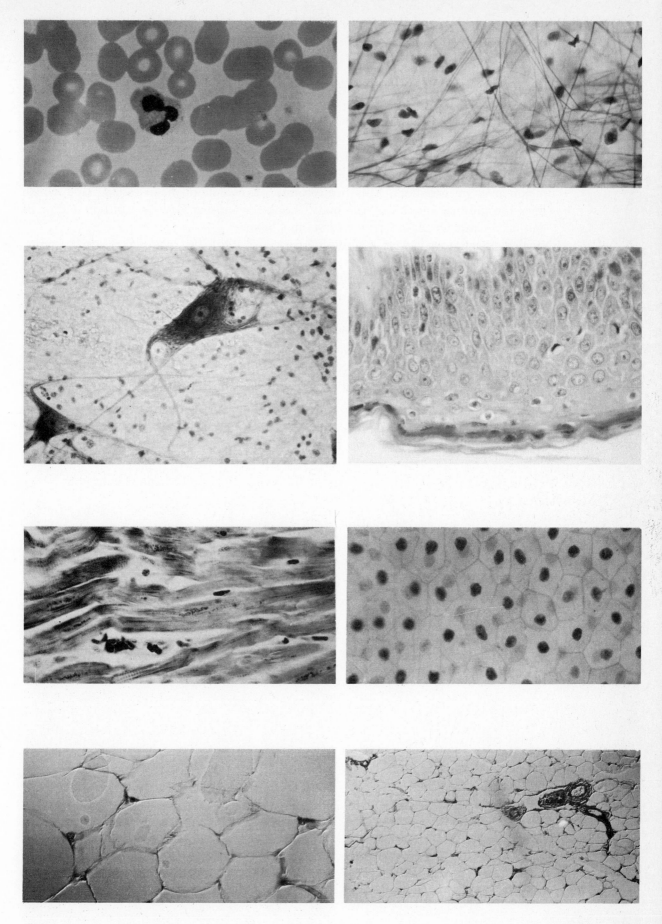

where they are almost invisible. A few nuclei can be seen in the high power view. The oval, dark-staining structure on the left in the low power photomicrograph is a tiny blood vessel. In this preparation the fat has been dissolved so that only the thin edge of cytoplasm, a few nuclei, and the cell membranes are visible. In order to make the cells clearer, special (Normarsky) optics were used, so these photomicrographs will be similar, but not identical, to your microscopic image.

Epithelial tissue (high power): *Simple squamous epithelium* (shown) consists of a thin layer, one cell thick, of hexagon-shaped cells. The nucleus stains darkly, and the cell membranes are indistinct. This tissue forms a thin, cellophane-like membrane, which lines the internal cavities of the body. Look for another view of a different type of epithelial tissue, under low power (described below).

Epithelial tissue (low power, longitudinal section): *Stratified epithelium*, such as comprises the skin, is more complex than simple squamous. It consists of a sequence of epithelial cell types, beginning furthest from the surface with columnar cells. These change to cuboidal cells which eventually flatten to squamous-type cells. At the surface of the skin, flattened epithelial cells die, become cornified (hardened), and are sloughed off. Note the flat layer of dead epithelial cells peeling off the skin surface.

Nervous tissue (high power): Two *neurons* (nerve cells) are shown. In the lower one the nucleus appears as a light circle containing a dark nucleolus. The cell body or cyton is roughly triangular in these neurons. Note the cytoplasmic extensions, long axons and shorter dendrites, which carry the nerve impulses.

ASSIGNMENT

Write an essay of about fifty words on the following topic, to be handed in at your next laboratory.

If cells did not develop their particular characteristics because the body needed them, what explanation is there for the facts that muscle cells *are* long and springy, blood cells *do* contain the respiratory pigment hemoglobin, and bone cells *do* produce a hard matrix capable of supporting the body? In other words, how can the origin of cell specialization be explained on a nonteleological basis?

Make sure you will be able to participate in a class discussion of teleology at your next laboratory meeting.

III Physical and Chemical Processes Within the Cell

8 The Model

Physical and Chemical Properties of the Protoplast
1: Osmosis

Our knowledge of the nature and processes of life has increased by leaps and bounds during the past several hundred years. In the eighteenth and nineteenth centuries most biologists contented themselves with objective observation of the biological environment. They examined their surroundings with a concern for detail, knowing that accurate description is the foundation of valid scientific thinking. Animals were described in books called *bestiaries;* plants were catalogued, classified, and minutely examined.

As biology became characterized by a broad foundation of facts compiled by these ardent naturalists, it became possible to suggest relationships among the facts and to develop means for testing these suggested relationships. It is important to realize that the formation of hypotheses cannot progress until a sufficient body of observations or facts exists. It should also be clear that as these hypotheses are tested, some will be shown to be valid, and new facts will then be added to our body of knowledge. By exposing relationships not evident before, valid hypotheses serve to give direction to scientific inquiry.

PRELIMINARY INFORMATION

Let us consider a hypothetical situation concerning the development of our knowledge about the cell. A scientist reads what others have observed concerning the cell. He learns that the cell carries on various processes associated with life and that it contains food molecules. These facts are at a relatively simple level; they are reported observations. He becomes inquisitive about where the food enters the cell. He reads what others have observed about the entrance of food into the cell and finds.

(1) One-celled animals sometimes have structures which collect the food, but it is stored in spaces in the cell and is separated from the cytoplasm until digested.

(2) Plant cells make their own food and store it in visible granules in the cytoplasm, but no one has seen the food pass out of one cell and into another.

(3) Cells of multicelled animals usually have no obvious way of obtaining food. They are usually specialized to perform such tasks as carrying oxygen or nerve impulses or to provide support and protection, and they have modifications to permit them to perform these functions. Most cells, however, do not have any specialized food-gathering organelles.

The scientist examines many cells and decides that the most logical place for the entrance of food into the cell is through that portion which is in contact with the surrounding environment, the cell membrane. He decides to make some preliminary observations to develop his idea. He places a drop of oil on a glass slide. The drop appears round and has a definite edge similar in outward appearance to a cell membrane. He sprinkles some chalk dust on the oil droplet and notes that after a period of time the dust has formed a ring at the base of the droplet. None of the dust has entered the drop. Somehow the outer boundary of the drop has prevented the entry of chalk dust particles! Obviously, the drop has the capacity to exclude particles. But food particles are large compared to molecules of water. Is it possible that the water molecules can pass through the cell membrane but larger particles are prevented from doing so? In other words, is it possible that the small size of the water molecules allows them to enter and leave the cell through the membrane more freely than the larger food molecules? If so, how does food actually enter the cell?

Having received some provisional confirmation of his idea that the size of particles and/or molecules affects their mode of entry into the cell from his preliminary observations, the scientists constructs a hypothesis.

□ Hypothesis: *The cell membrane is the region which permits the entry of water into the cell, but it prevents larger particles, such as food, from entering.*

In order to test this hypothesis, the scientist must find out more about the characteristics of the membrane. Since the cell membrane is so tenuous and the cell itself is so small, the scientist casts about to find a way of duplicating the aspects of the cell he is studying on a size level which will permit experimentation. He decides to devise a model of the cell.

THE MODEL

Often scientists find it helpful to construct a model, a simplified approximation of some object or phenomenon, in order to consider it more clearly. Our concept of the atom is an example of a model. The atomic scientist uses the "planetary" or Bohr model of the atom because from it he can predict certain relationships, yet he understands that the atom is more complex than his model. A model imitates particular aspects of an object or phenomenon. The scientist is able to study each of these aspects independently and thus avoid the likelihood of confusion that might arise from the complexity of the original.

Our model of the cell will consist of a cellophane sausage-casing containing a saturated salt solution. This "cell" is to be placed in water to imitate the relationship between the cell and the lymph which surrounds it in the human body. Observe the "cell" in its water bath now and again in 1 hour. As a comparison, place another sausage-casing cell, filled with distilled water, in the same dish. Observe also another, even simpler model, the osmometer setup at the front table. It functions like your model cell and can be used to check the validity of your other observations. **SET UP YOUR MODEL CELL NOW.**

THE OSMOMETER

The osmometer at the front table behaves in the same manner as your model cell since its lower opening is covered with a membrane of the same material which constitutes the membrane of the model cell. Record in Table 8–1 the distance the water rises or falls every 15 minutes for the next hour and a quarter.

TABLE 8-1   Changes in Level of Salt Solution in Osmometer

Reading	Time	Height
(now) 0		
1		
2		
3		
4		

1. Which direction did the fluid in the osmometer move (up, down, stayed the same)?

_____

Compare the distances the fluid moved up the tube between readings 1 and 2 and between

readings 3 and 4 _____

Was there a difference in the distances moved? _____

State a hypothesis which might explain why this difference occurred.

The following questions are to be answered when you have completed your observations of the model cell and osmometer. In the meantime proceed to the next section, "The Movement of Particles and Molecules."

2. What is the difference between the contents of our model cells and the water sur-

rounding them? _____

_____

_____

3. Does the cell behave in the same manner in the body as our model does? What possible difference can there be between the functioning of the real cell membrane and the cellophane

membrane? _____

_____

4. In what ways are they the same in function? _____

_____

5. What has happened to your model cells? _____

_____

6. What might eventually happen to your model cells if allowed to remain in the water?

_____

7. What can you deduce from your answer concerning the relationship between your

body cells and your blood? _____

_____

_____

_____

8. Can you, at this point, make a definite statement about whether or not *large* particles

(such as undigested food molecules) enter the cell through the cell membrane? _____

_____

_____

The Movement of Particles and Molecules

By reading about digestion, our scientist finds out that its function is to break down food into relatively small particles (or molecules).

Since he has suggested that the particles of food cannot pass through the cell membrane, it will be necessary to find out something about the particles themselves.

Place a very small drop of India ink in a drop of water on a slide. Mix the two. The dilution is correct if the mixture appears light grey when the slide is placed on a piece of white paper. Place a cover slip on the drop and examine it first under *low power* and then under *high power.*

Observe carefully *one* particle under high power for several seconds. A particle is composed of many molecules. (*Note:* The flowing movement of all the particles in one direction is a result of the fact that the stage of your microscope is not perfectly level. This is not what you are looking for. You will find another sort of movement superimposed on the flowing motion if you concentrate on one particle and ignore its constant movement in one direction.)

9. Record your observations. _____

_____

10. On the basis of the foregoing observations, is it absolutely necessary that the cell

membrane supply energy in order to bring molecules or particles into the cell? _____

Explain clearly. _____

_____

_____

11. Formulate a hypothesis as to how small molecules such as water molecules enter the cell. Assume, for this purpose, that the cell membrane does not expend energy (does not do

anything when the particles pass through it). _____

_____

_____

_____

12. How does *chance* enter into this hypothesis?_____

_____

_____

13. Explain, as well as you can at this point, why the contents of the cell (except water)

do not leave the cell in the same way that water enters.* _____

_____

Osmosis

You have seen how very small **particles** (a particle is composed of many molecules) of India ink have a characteristic random movement. The ink particles themselves are relatively large when compared to individual **molecules.** The movement of the ink particles which we saw under the microscope was caused by **collisions with rapidly moving water molecules** which were too small to see. Thus the most important interpretation of the ink and water observation is that molecules of water are in constant motion.

*If you are not sure of your answer, read the next section, "Osmosis," and then come back to this question.

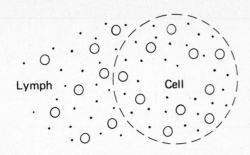

FIGURE 8-1   Cell surrounded by lymph. Water molecules can enter and leave cells through holes in the membrane. Other molecules are too large.

When a cell is bathed in the watery fluid called lymph (as are all living cells), the contents of the cell are separated from the lymph by the cell membrane. The cell membrane permits the water molecules to pass freely through *in both directions,* but prohibits larger particles from either entering or leaving. Such a membrane, which permits some objects to pass through but not others (depending on their *size*) is said to be **semipermeable.** Thus we can represent the cell membrane as shown in Figure 8-1. The dots represent water molecules. The circles represent the larger molecules found in both the lymph and the protoplasm of the cell. Note that the water molecules can move freely through the cell membrane but the other molecules are too large to pass through the pores in the membrane. At any given moment a number of *randomly* moving water molecules will be passing through the membrane from the cell to the outside and, at the same time, some of the water molecules on the outside will be passing in (see Figure 8-2).

The number of molecules passing through in each direction is determined primarily by the quantity of molecules on each side of the membrane. Thus if there are 5000 molecules of water outside the cell and only 500 molecules inside, it is reasonable to assume that there will be fewer water molecules moving out of the cell and more moving in. In Figure 8-3 there is a net gain of 500 molecules of water into the cell through the membrane.

Cells and lymph contain more than just pure water. They are both comprised of suspensions and solutions of relatively large particles and molecules in the water. When there are many particles and large molecules relative to the numbers of molecules of water, the solution is said to be **concentrated.** When there are few particles and many water molecules, the solution is **dilute.** A solution having the same ratio of water molecules to particles found inside the cell is said to be **isotonic** to the cell. The process by which **water molecules** pass freely through a **semipermeable** membrane is called **osmosis.**

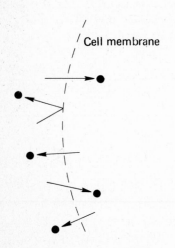

FIGURE 8-2   **Water molecules passing through membrane.**

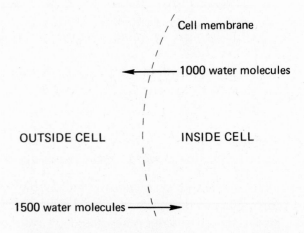

OUTSIDE CELL   INSIDE CELL

FIGURE 8-3   **A cell placed in water.** Note that more water molecules are entering the cell than are leaving it.

14. Does your understanding of osmosis permit you to explain how relatively large molecules such as amino acids and sugar enter the cell?_____ Explain._____

_____

15. Is it likely that the cell membrane has a passive role when it comes to the entry of large molecules?_____ Why or why not?_____

_____

Examine your model cell and the osmometer. It is likely that enough time has elapsed to make possible valid observations. Go back to questions 1b–8 and answer them now. Then return to question 16.

16. At this point you (and our hypothetical scientist) have gathered a considerable amount of information regarding the functions of the cell membrane. Summarize this information in the space below.

17. Are you satisfied that the scientist now has a positive solution to his problem? Has he fully tested the hypothesis "The cell membrane is the region which permits the entrance of

water molecules into the cell"? _____

In order to be sure that he is correct, the scientist can test the predictive validity of his conclusion. One of the ways of testing a tentatively proven hypothesis is to determine whether or not predictions based on the hypothesis turn out to be true.

18. Does this method, *in itself,* supply *definite proof* that the hypothesis is correct?_____

Explain. _____

_____

■ THE PROBLEM: *To test the hypothesis that the cell membrane permits the entrance of water molecules into the cell.*

In order to test the hypothesis that the cell membrane is involved in allowing water to enter and/or leave the cell, you will make predictions about the appearance of red blood cells in various concentrations of saltwater (saline solutions). Make your predictions in the spaces provided, and beneath each prediction draw the red blood cells as you observe them in each solution. Revise or corroborate your hypothesis on the basis of these observation. **Be sure to make your predictions before you examine the slides.**

Observations

A. Place a drop of blood in a solution in which the ratio of salt molecules to water molecules is *the same* as the ratio of salt, protein, etc., molecules to water molecules in the cell. This is known as an *isotonic* salt solution: 0.9% (Read B and C to clarify the foregoing.)

B. Place a drop of blood in a solution in which the ratio of salt molecules to water molecules is *less* than the ratio of salt (protein, etc.) molecules to water molecules in the cell (a *dilute* salt solution [we will use distilled water to emphasize the effect of this relationship]).

C. Place a drop of blood in a solution in which the ratio of salt molecules to water molecules is *greater* than the ratio of salt (protein, etc.) molecules to water molecules in the cell (a *concentrated* salt solution).

19. Prediction for solution A (your idea of what should happen to the cells in an isotonic

salt solution) _____

_____

Drawings of Cells in Solution A

Prediction   *Observation*

20. Prediction for solution B _____

_____

Drawings of Cells in Solution B

Prediction   *Observation*

21. Prediction for solution C _____

_____

Drawings of Cells in Solution C

Prediction   *Observation*

22. Do your drawings corroborate your predictions? _____
If not, modify your predictions in light of the new data and test them again. (*Note:* Your instructor may have made available *Elodea* leaves at the front table. If so, make identical preparations to those of your blood slides and examine them to confirm your blood observations. The rigid cell wall of the *Elodea* will provide a frame of reference lacking in blood cells.)

23. Your predictions concern the passage of *water molecules* through the cell membrane. Make up a general statement (based on your data) about the manner in which water enters or

leaves the cell. _____

_____

_____

_____

_____

_____

_____

_____

_____

_____

_____

24. Is your generalization also adequate to explain the entry of the relatively large food

molecules? _____

REVIEW QUESTIONS

Review the entire exercise before answering the following questions.

1. What evidence have you that food molecules can or cannot *(circle one)* enter the cell in the same manner as water?

2. Can the cell membrane remain passive (as it does when water passes through it) when

relatively large molecules of food enter?_____What hypothesis can account for the entry of food molecules?

3. What is a model? What is its purpose? Why did we consider the "sausage-casing" bag a model of a cell?

4. Learn the definitions of the following terms by looking them up in reference books in the library. Make sure you are able to use each term to explain the appropriate observation made during this laboratory period.

        a. Hypotonic
        b. Hypertonic
        c. Isotonic

       d. Crenation (referring to cells, not to leaves)
       e. Brownian movement
       f. Diffusion
       g. Hemolysis
       h. Turgor
       i. Plasmolysis

Note: Biological investigations are more and more in the realm of particles, atoms, and molecules. Thus biochemistry and biophysics seem to hold the answers to some of the most fundamental biological questions.

9 Relationship Between Facts and Formulation of Hypotheses

Physical and Chemical Properties of the Protoplast 2: Chlorophyll and Photosynthesis

A hypothesis is a suggested relationship between facts. In order to construct a potentially valid hypothesis it is necessary to study a portion of the environment until relationships become apparent. This implies that a considerable amount of effort must be expended toward fact gathering before any facility with hypothesis construction can be achieved. Every science is constantly undergoing a "fact accumulation" phase which will be the foundation for further development of the discipline.

In order to solve today's problem, it will be necessary for you to use some knowledge recently obtained by chemists and biophysicists, augmented by your own observations. *Only when all of this information is correlated will it be possible to make up meaningful hypotheses leading to the solution of the problem.*

PRELIMINARY INFORMATION

All living things rely on the capacity of plants to harness the sun's energy in a food-manufacturing process called photosynthesis. No animal, not excepting man, can put together such simple compounds as water and carbon dioxide and construct food molecules from them on a large scale.

The overall process of photosynthesis is represented by the following equation.

$$12\ H_2O + 6\ CO_2 + \text{light energy} \xrightarrow{\text{chlorophyll}} \underset{\text{Glucose}}{C_6H_{12}O_6} + \underset{\text{Oxygen}}{6\ O_2} + \underset{\text{Water}}{6\ H_2O}$$

Glossary

ATP: Adenosine triphosphate is a molecule used for energy storage. When it releases a phosphate to become adenosine diphosphate, a packet of energy is made available for energy-requiring reactions.

NADPH$_2$: Nicotinamide adenine dinucleotide phosphate, reduced form, is a source of energy and electrons. It has about seven times as much useful energy as ATP. When the energy and the electrons are released, NADPH$_2$ becomes NADP$^+$.

RuDP: Ribulose diphosphate is a five-carbon sugar that combines with CO_2 to form an intermediate six-carbon compound which immediately breaks down into two PGA molecules.

PGA: Phosphoglyceric acid is a low energy three-carbon acid which is converted to PGAL by the application of energy and the addition of electrons.

PGAL: Phosphoglyceraldehyde is a three-carbon sugar which can be joined with other PGAL molecules to become glucose. Many RuDP molecules are left over. These are recycled, eventually being used to produce more glucose molecules.

In photosynthesis, sunlight supplies energy which, in the chloroplast of the plant cell, converts carbon dioxide to sugar and other important molecules which sustain life. The sun's energy is eventually converted into two energy-storying molecules, ATP and $NADPH_2$. They supply the energy to convert PGA into PGAL. **The PGAL becomes glucose and other food molecules.** Thus the carbon dioxide and RuDP, a five-carbon sugar, supply the necessary atoms to construct the PGAL and, eventually, the glucose, while the ATP and $NADPH_2$ supply the energy.

The reaction is summarized below.

Light Reaction

In a reaction which can occur only in the presence of light, two hydrogen atoms and electrons are removed from a water molecule, leaving an oxygen atom which combines with another freed oxygen atom; they escape as gaseous oxygen (O_2). Since the hydrogen of the H_2O attaches tenaciously to the oxygen, energy is required to break the bond. Light may be thought of as streams of tiny packets of energy called photons. The photons are absorbed by special chlorophyll–protein molecules. When enough energy is transferred to one of these molecules, some of its electrons become so excited that they pop out of the molecules in pairs. The energy of these electrons is used to form energy storage molecules or energy carriers, ATP and $NADPH_2$ (the electrons themselves help form the $NADPH_2$).

After a number of reactions involving released electrons, two ATP energy carrier molecules are formed. The electrons are excited once again and pass through another chain of molecules. This time the two electrons combine with an $NADP^+$ molecule and with two hydrogen ions (left over from the splitting of water) to form $NADPH_2$ (an electron-carrying molecule). Thus the energy carrier, ATP, and the energy and electron carrier, $NADPH_2$, are produced in the light phase.

Synthetic Reaction

Next occurs a reaction which uses the energy-rich molecules produced in the light phase. This reaction does not need light since the energy carriers from the light reactions are present.

1. An enzyme hooks up a carbon dioxide (CO_2) molecule to a five-carbon RuDP. This reaction, in which CO_2 is removed from air and combined into a carbon chain, is called carbon fixation. The RuDP + CO_2 forms an unstable six-carbon molecule which immediately breaks down into two three-carbon PGA molecules. For every six CO_2 molecules fixed, 12 PGA molecules are produced.

2. The 12 PGA molecules (three-carbon acids) now enter a series of reactions resulting in the formation of three-carbon sugars called PGAL. This requires great quantities of energy and many electrons. No less than 12 ATP and 12 $NADPH_2$ molecules from the light reactions are involved.

3. The PGAL molecules are restructured to form the glucose molecule, using up six more ATP molecules. There are some leftover pieces of the PGAL molecules which are reassembled

into more RuDP molecules, which replace the RuDP molecules used up in the beginning of the synthetic phase reaction.

4. The spent energy molecules, ADP (the used-up remainder of ATP) and $NADP^+$ (the used-up $NADPH_2$), are returned to the light reaction sites where they are converted once again into $NADPH_2$ and ATP. Thus the cycle is self-replenishing except for energy and CO_2.

If we examine the photosynthetic reaction again *in toto*, we can see the origin of each product.

$$12 H_2O \quad + \quad 6 CO_2 \quad + E \longrightarrow \quad C_6H_{12}O_6 \quad + \quad 6 O_2 \quad + \quad 6 H_2O$$

Water molecules contribute electrons and atoms to manufacture of ATP, $NADPH_2$, and O_2 (oxygen gas), which is released as a by-product.	Only 1 molecule of CO_2 is used in synthesis of each glucose molecule. The remaining 5 are used to replenish RuDP.	Glucose is derived from PGAL restructured through expenditure of energy (ATP) and electrons ($NADPH_2$).	Oxygen is produced when light energy, converted by chlorophyll molecules, breaks up water molecules.	Water is a by-product released during the synthetic phase. Not the same as original water molecules at beginning of equation.

History of the Problem

When taxonomists considered dividing all life into five kingdoms, rather than two, one proposal suggested grouping together those living things which contained a compound called cellulose in their cell walls as Kingdom Plantae. A problem arose when cellulose was found in fungi, and, to the taxonomists' shock, even in the walls of bag-like *animals* called tunicates (subphylum Urochordata) which spend their lives straining water to extract microscopic food. Since the presence or absence of cellulose seemed to be too broad a criterion, other characteristics were searched for. The presence or absence of chlorophyll, the green pigment that carries on photosynthesis, was proposed as a good dividing line for Kingdom Plantae. Critics rejected the idea because many plants are red, yellow-brown, and other colors and might therefore possess pigments other than chlorophyll. In today's exercise you will have a chance to determine whether or not the presence of chlorophyll is an adequate criterion for the establishment of Kingdom Plantae.

■ THE PROBLEM: *Can the presence of chlorophyll be used to differentiate Kingdom Plantae from the other kingdoms?*

1. List some plants or plant-like protists which are not green.

a. _____   d. _____

b. _____   e. _____

c. _____

At the front table you will find representatives of some nongreen plants and fungi. You will notice three conspicuous groupings which seem not to contain chlorophyll. They are represented by *Fucus* and other red and brown seaweeds; red maples, red cabbage, and other red tracheophytes; and mushrooms and other fungi.

2. Hypothesize as to how the above organisms obtain nourishment if they are not green. Write several possible answers.

a._____

b._____

c._____

d._____

So far, all you are *sure of* is that there are numerous organisms which have cellulose cell walls and lack green color. Below will be found directions for performing exercises designed to give you further information, so that you can develop hypotheses which may effectively lead you to the solution of your problem.

Observation 1

Each group of four students will grind up some spinach or ivy leaves with a mortar and pestle and a little clean sand. Add from time to time 1 ml of acetone until about 5 ml are in the mortar. Take a pipetful of the green extract and put ten drops one on top of the other at the place on the strip of filter paper indicated in Figure 9-1. Allow each of the green drops to dry before adding the next. Set up your apparatus according to the figure, pouring 1 ml of petroleum ether–acetone solvent into the test tube first.

Petroleum ether vapor is explosively flammable! No flames or smoking in the room.

Allow the solvent to be in contact with the filter paper until it has moved up to ½ inch

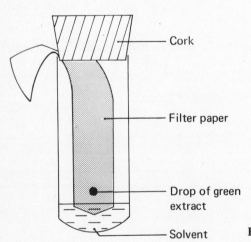

Cork

Filter paper

Drop of green extract

Solvent

FIGURE 9-1 Paper chromatography apparatus.

below the cork (15–25 minutes). Then place the strip of paper under a bright light and examine it.

The molecules of chlorophyll, like all molecules, are constantly in motion. Since the chlorophyll is soluble in the acetone–ether solvent, the two solutions will mix and the molecules of chlorophyll and other substances of a similar nature will tend to disperse in the semiliquid medium of the wet filter paper. Several poorly understood physical phenomena are involved in governing the rate of flow of the chlorophyll up the paper strip. Certainly such factors as size and weight of a molecule, as well as degree of ionization, affect its rate of migration. The result of this process (paper chromatography) is to separate complex groups of compounds into their components, according to the rate of migration of each kind of molecule. This technique, while simple to perform, is a highly sophisticated research tool used to uncover the composition of many biologically important substances. It is especially useful in differentiating and identifying the amino acids in various proteins.

3. Record your results. _____

_____

_____

_____

_____

It is important for you to realize that you have *not* performed an experiment. You have followed the directions of this laboratory exercise to help you obtain data pertinent to the solution of your problem.

4. Why is the foregoing an exercise and not an experiment? _____

_____

_____

5. Summarize your observations. Does the green extract contain only one pigment? Is

chlorophyll the only pigment present? (Remember chlorophyll is green.) _____

_____

_____

_____

6. Why are these observations and not conclusions? _____

_____

_____

Does the fact that certain pigments can mask (hide) the appearance of other pigments suggest a hypothesis to you? If so, state it.

☐ Hypothesis: _____

_____

_____

The experiment outlined below will allow you to test your hypothesis concerning the masking of pigments.

Experiment

At the front table you will find beakers and red cabbage or *Coleus* leaves. Take a leaf and boil it in a beaker half full of water.

7. What do you think happens to cells and cell walls during the boiling? _____

_____

8. What are your observations?

    a. Color of water _____

    b. Color of leaf _____

9. What are the implications of your observations? _____

_____

_____

10. Is your hypothesis valid in this case? _____

11. Do you now have a solution to the problem? _____

12. Does this mean that other hypotheses are automatically invalid and should not

be tested? _____ Why or why not? _____

_____

_____

_____

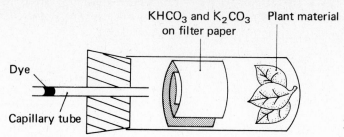

KHCO$_3$ and K$_2$CO$_3$ Plant material
on filter paper

Dye

Capillary tube

FIGURE 9-2   Manometer.

Observation 2

Examine Figure 9-2. A manometer is a device which measures the pressure of the gases in a vial. In this case the gas pressures will be affected by (a) the metabolism of the living organisms in the vials and (b) the carbon dioxide liberated by the potassium carbonate compounds on the filter paper inside the vials. These carbonate compounds will react to produce a *fixed* relatively high carbon dioxide atmosphere in the chamber. *No matter how the living organism inside the manometer produces or utilizes carbon dioxide, the partial pressure of this gas in the capillary tube will remain constant.*

13. As you observe the functioning manometer, you will see the drop of dye move within the capillary tube. If the carbon dioxide partial pressure is kept constant by the

carbonate compounds, what gas is causing the drop of dye to move? _____
(If you do not know the answer to this question, consult the photosynthesis equation on page 107.)

14. Temperature is a variable in this experiment because heat causes gases to expand and may affect the position of the drop of dye. Read the directions in the paragraph immediately

following. How is the temperature variable controlled? _____

_____

_____

_____

_____

Place a folded 2-inch leaf from a green plant in one manometer and a piece of fresh (live) mushroom in the other. Dip one end of a ¼-inch-wide strip of filter paper in the beaker of KHCO$_3$ and the other end in the K$_2$CO$_3$. Loosely coil the paper and place it in one of the manometers. Do the same for the other manometer. Make sure that the saturated filter paper does not touch the live plant material. Dip the end of each capillary tube into red ink and tap until the drop is near the cork. Press on the corks in each vial, compressing the air inside until you have matched the bottom of one drop of dye with the other. Draw a line on a piece of white paper and place both vials on their sides so that the bottom of each drop of dye is at the same position on the "starting line." After 10 minutes measure the distance, in millimeters, from the starting line to the position of the drop of dye. Continue to take readings every 3 minutes on each system.

15. Record your data below for 10 readings. Be sure to set up your apparatus near the window, or under a bright light, for maximum photosynthesis.

_____   _____   _____   _____   _____

_____   _____   _____   _____   _____

16. Summarize your results: _____

_____

_____

_____

17. Does this information relate to your hypothesis? Explain. _____

_____

_____

18. What do you find to be the relationship between the amount of preliminary information and the formulation of hypotheses? _____

_____

_____

19. In the space below, state the problem. Then state your solution to the problem, carefully giving proof you are correct.

20. Have you proved that some plants and/or fungi do or do not carry on photosynthesis?_____ Explain._____

_____

_____

_____

21. State your hypothesis. _____

_____

_____

_____

22. Describe the test of your hypothesis. _____

_____

_____

23. Was your hypothesis valid or not?_____ Why or why not?_____

_____

_____

IV · Origin of Life

10 Development of Biological Theory

Origin of Life 1: Spontaneous Generation

The development of knowledge in biology has progressed from simple observation to more complex understanding of phenomena in a kind of geometric progression. A culmination of this development has been reached within our lifetime. In the first fifty years of the twentieth century more knowledge has been accumulated than in all of man's history before that time. How can we explain this sudden expansion of our knowledge?

One possible explanation is that we have suddenly become more intelligent. This hypothesis does not seem too promising. Is it a sign of increased intelligence that we are still the victims of our own self-destructive wars?

Another possible explanation lies in the nature of theory. A **theory** is the accumulation of a number of related, **validated hypotheses**. These hypotheses were suggested by **facts** (observations). As more and more hypotheses are tested by experiments and are found to be valid, our store of usable information increases. Finally, a particularly perceptive person is able to see a relationship among several of these hypotheses, and a theory is born. A theory, while similar to a hypothesis in that both are suggested interrelationships among facts, differs from the hypothesis in the magnitude of its scope. Many people were aware that objects fall "downward" if not obstructed, but only Newton was able to show the profound implications of this phenomenon in relation to the behavior of the planets and indeed of all matter. Newton formulated a theory of gravitation. After several hundred years of testing this theory for its universality, scientists were willing to accept it as a law. This meant that when dealing with any freely falling object one could predict that it would fall, at a certain rate, toward the gravitational center of the most massive object in its vicinity. (Knowledge is never absolute, and we find interesting variations from the law of gravity in nuclear physics.)

A theory, then, is really an accumulation of interrelated facts (that is, hypotheses which have been proven true and thus have the status of fact). But theory has an additional value. By pulling together these facts and showing us how they interact, theory allows us to manipulate them more easily. By showing that the planets, like any other object, respond to gravitation, Newton made it possible for others to solve the riddle of planetary movement. In effect, theory suggests new hypotheses which in turn will be incorporated into new theory. Thus facts suggest theory and theory suggests new facts.

In this feedback relationship we have a possible explanation for our ever-increasing tempo of discovery. Once Columbus' theory that there was land on the other side of the ocean was shown to be valid, numerous other explorers set sail toward the New World, and many new areas were discovered. Figure 10–1 is a pictorial representation of the geometric nature of the accumulation of knowledge.

PRELIMINARY INFORMATION

In this laboratory we will examine the development of the theory of spontaneous generation from its inception in antiquity, through its rejection in the nineteenth century, to its resurrection in the twentieth century.

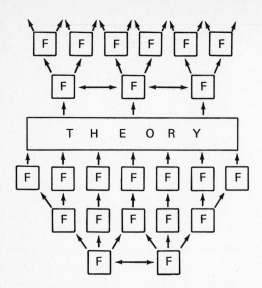

FIGURE 10-1   The relationship between fact and theory.

Students seem to find it hard to believe that some of the wisest of the ancient, medieval, and relatively recent (mid-nineteenth century) scientists and philosophers—Aristotle, among others—believed that certain forms of life originated from nonliving substances. Yet this concept, called **spontaneous generation**, had many adherents through the centuries; in fact, it was practically undisputed until the sixteenth century, and vestiges of its popular acceptance can still be found in such statements as "Don't give sugar to a dog; it will give him worms" and "Dust breeds germs."

The readings presented below should help you to understand that some of the ideas which you take for granted were developed through centuries of debate. As you read each selection, try to understand the level of knowledge on which the conclusions were based. You will see that it was more logical for people of the ancient world to believe in spontaneous generation than not to, since much of the information available at the time seemed to point to the existence of this phenomenon.

Read the following selections, noticing the date when each was written.

From **Nature of Man**—*Hippocrates* (about 400 B.C.)

This statement by the "father of medicine" is included to exemplify the more liberal and rational reasoning of his day. Note the impressive logic behind what is said and how the argument conforms to experience. After all, is it not true that phlegm appears in largest quantity in winter? Yellow bile was supposed to be produced in the liver and black bile in the spleen. While not directly concerned with spontaneous generation, this reading helps us to understand how observations lead to such theories.

For I do not say that a man is air, or fire, or water, or earth, or anything else that is not an obvious constituent of man; such accounts I leave to those who care to give them. Those, however, who give them have not in my opinion correct knowledge. . . .

Now about these men I have said enough, and I will turn to physicians. Some of them say that man is blood, others that he is bile, a few of them that he is phlegm. . . . The body of man has in itself blood, phlegm, yellow bile and black bile. These make up

*Hippocrates, Volume IV, translated by W. H. S. Jones, Harvard University Press, 1931.

the nature of his body, and through these he feels pain or enjoys health. Now he enjoys the most perfect health when these elements are duly proportioned to one another in respect of compounding, power, bulk, and when they are perfectly mingled. Pain is felt when one of these elements is in defect or excess, or is isolated in the body without being compounded with all the others. For when an element is isolated and stands by itself, not only must the place where it left become diseased, but the place where it stands in a flood must, because of the excess, cause pain and distress. In fact when more of an element flows out of the body than is necessary to get rid of superfluity, the emptying causes pain.

... Phlegm increases in a man in winter; for phlegm, being the coldest constituent of the body, is closest akin to winter. It is the winter when sputum and the nasal discharge of men is fullest of phlegm; at this season mostly swellings become white and diseases generally phlegmatic. And in spring too, phlegm still remains strongly in the body while the blood increases. And in summer blood is still strong, and bile rises in the body and extends until autumn. In the summer phlegm is at its weakest. But in autumn, blood becomes weakest in man, for autumn is dry and begins from this point to chill him. It is black bile which in autumn is strongest. All these elements are then always comprised in the body of man, but as the year goes round they become now greater now less, each in turn and according to its nature.

The Production of Living Bees from the Carcass of a Dead Calf
(*From* **Georgics**)—*Virgil* (29 B.C.)*

In the first place, a spot small and confined for the very purpose is chosen out; this they close in with narrow roof-tiling, and straitened walls, and insert four windows, with slanting light from the four quarters of heaven. Next they look out for a bull-calf, whose horns have begun to arch over a brow that has seen two summers; they take him and seal up his two nostrils and his mouth's breath, spite of furious struggles; and after he has been slain by their blows, his flesh through his unbroken hide is beaten to a jelly. As he lies, they leave him in his barred prison, placing under his ribs broken bits of bough, and thyme; and fresh-plucked cassia. This goes on when the west winds first play upon the waters, ere the meadows are empurpled with fresh spring-tide hues, ere the chattering swallow hangs her nest from the rafters. Meanwhile, in his softened bones the sap has been heated and begins to ferment, and living things of strange manner to look upon, at first with no feet to crawl on, but soon even with wings to buzz and fly with, swarm confusedly, and skim the empty air more and more, till, like a burst of rain from a summer cloud, out they break, or like arrows from the rebounding cord, whenever the light-armed Parthian strikes up the prelude of battle.

From **De Generatione Animalium (On the Generation of Animals)**—
Aristotle (Fourth Century B.C.)†

Now some animals come into being from the union of male and female, i.e., all those kinds of animals which possess the two sexes. This is not the case with all of them; though in the sanguinae (animals with blood) with few exceptions the creature when its growth is complete, is either male or female, and though some bloodless animals have sexes so that they generate offspring of the same kind, yet other bloodless animals generate indeed, but not from offsping of the same kind; such are all that come into being not from a union of the sexes, but from decaying earth and excrements. To speak generally, if we take all animals which change their locality, some by swimming, others by flying, others by walking, we find in these two sexes, not only in the sanguinae but also in some of the bloodless animals; and this applies in the case of the latter sometimes to the whole class, as in the cephalopoda and crustacea, but in the case of the insects only to the majority. Of these, all which are produced by union of animals of the same kind generate also from their own kind, but all of which are not produced by animals, but from decaying matter, generate indeed, but produce another kind, and the offspring is neither male nor female; such are some of the insects.

But all those creatures which do not move, as the testacea and animals that live by clinging to something else, inasmuch as their nature resembles that of

The Works of Virgil, translated by John Conington and Willard Small, Lee and Shepard, Boston, 1892, pp. 109–10.

†*The Basic Works of Aristotle,* Richard McKeon, editor, Random House, New York, 1941, pp. 665-66.

plants, have no sex any more than plants have, but as applied to them the word is used in virtue of a similarity and analogy. For there is a slight distinction of this sort, since even in plants we find in the same kind some trees which bear fruit and others which, while bearing none themselves, yet contribute to the ripening of the fruits of those which do, as in the case of the fig-tree and the caprifig.

The same holds good also in plants, some coming into being from seed, and others, as it were, by the spontaneous action of Nature, arising either from decomposition of the earth or of some parts in other plants, for some are not formed by themselves separately but are produced upon other trees, as in the mistletoe.

From **Essay on Spontaneous Generation**—*J. B. van Helmont*
(Seventeenth Century A.D.)*

Thus it comes about that lice, bugs, bees and worms become our guests and the companions of our afflictions, and are as it were born of our inmost parts and our excrements; futhermore, if a dirty undergarment is squeezed into the mouth of a vessel containing wheat, within a few (say 21) days the ferment drained from the garment and transformed by the smell of the grain, encrusts the wheat itself with its own skin and turns it into mice. . . and what is more remarkable, the mice from corn and undergarments are neither weanlings, nor sucklings, nor premature, but they jump out fully formed.

From **The Origin of Life**—*A. I. Oparin* (A.D. 1923)†

The bodies of all animals, plants, and microbes are to a large extent composed of organic substances. Without them life is unthinkable. Therefore the primary formation of these substances, the formation of that basic material which eventually led to the appearance of all living creatures, necessarily marks the initial stage of the origin of life.

Organic substances differ from substances of the inorganic world, first and foremost, in that they are built around the element of carbon. This is easily demonstrated by heating various materials of animal and plant origin to a high temperature. All these materials burn when they are thus heated in air or are carbonized when air is shut out, while materials of inorganic nature—stone, glass, metal, do not carbonize no matter how long they are heated.

In organic substances carbon is compounded with other elements: hydrogen and oxygen (these two elements are contained in water), nitrogen, considerable quantities of which are contained in air, sulphur, phosphorus, and so forth. Various organic substances present extremely diversified compounds of these elements, but carbon invariably forms their base. Hydrocarbons—compounds of carbon and hydrogen—are the most elementary organic substances. Natural oil and such of its products as benzine, kerosene, etc., are mixtures of various hydrocarbons. On the basis of the latter chemists easily produce by synthesis numerous organic compounds, at times highly complex ones, possessing, in a number of cases, qualities akin to those of organic compounds extracted from living beings, as for example, sugar, fats, essential oils, etc.

In what way, then, did these organic substances originally appear on our planet?

When I first turned to the study of the problems of life's origin—some thirty years ago—the question of the primary formation of organic substances seemed a mystery which defied investigation and understanding. For direct observations of our natural environment showed that the vast mass of the organic substances of the animate world now originates on the Earth as a result of the vital activity of organisms. Living green plants, absorbing inorganic carbon from the air, produce from it with the use of solar energy the organic substances they require. Animals, fungi, bacteria, and other organisms which are not colored green obtain the necessary organic substances by feeding on plants or by decomposing their bodies and remains. In this manner, the entire modern living world owes its existence to this process of photosynthesis or the analogous process of chemosynthesis. Moreover, even all the organic substances which are deposited in the bowels of the earth in the form of peat, coal, and oil also for the most part originated as the result of the vital activity of numerous organisms that once inhabited the Earth and were later interred in the Earth's crust.

*J. B. van Helmont, *Ortus medicinae,* Amsterdam, 1648.
†A. I. Oparin, *The Origin of Life,* Foreign Languages Publishing House, Moscow, 1955, pp. 21-23.

Late in the previous century and early in this one, many scientists inferred from the aforesaid thought that organic substances on our planet can originate only biogenetically, that is, with the participation of organisms. This opinion which was predominant in science some thirty years ago set up serious problems in the way of solving the problem of life's origin. A vicious circle seemed to have arisen. In order to trace life to its sources it was necessary to know how organic substances are formed, but those latter, they alleged, were synthesized by living organisms alone. But that happens only when we restrict our observations to the confines of our planet. When we go further, we see that on a number of celestial bodies in our stellar universe organic substances are being formed abiogenetically, that is, in conditions which absolutely preclude the possibility of any organisms existing there.

With the development of science in the Renaissance period, men began to question the beliefs passed on undisputed from Aristotle's time. They sought evidence. This caused a gradual undermining of the theory of spontaneous generation, as observers noted that while bonfires did liberate sparks which *looked* like fireflies, no one had actually seen a spark turn into a firefly. In 1668, Francesco Redi published his *Observations on the Generation of Insects*. This experiment threatened to deal a mortal blow to the theory of spontaneous generation because it showed that flies are not derived spontaneously from dead meat. However, those persons committed to the theory took advantage of a then-recent observation made by Leeuwenhoeck, the first microscopist. By focusing his newly discovered microscope on a drop of pond water, Leeuwenhoeck was able to demonstrate the existence of myriads of heretofore invisible organisms. Eager proponents of spontaneous generation were quick to conclude that these were born of water or air, although Leeuwenhoeck, using his microscope, was able to trace the life cycles of several very small insects, demonstrating that they were *not* spontaneously generated. It was all too easy for believers in spontaneous generation to switch from mice and flies to bacteria and protozoans.

Thus the second phase of the battle of spontaneous generation began. Among the conflicts in this series of scientific arguments were those between Lazzaro Spallanzani (1729-99) and John Needham (1713-81) and between Louis Pasteur (1822-95) and F. A. Pouchet (1800-72).

Needham Versus Spallanzani

Needham, in 1749, published some experiments which, he stated, demonstrated that spontaneous generation existed. He put some mutton gravy in a flask, heated it, and sealed it with a cork. After several days the gravy in the closed flask had bacteria in it.

Spallanzani read of Needham's work and repeated it, sealing the flasks by melting closed their mouths and boiling them for several hours. He found that after cooling, no matter how long he waited, when he broke open the flasks there were no bacteria in them.

Needham concluded from Spallanzani's work that the prolonged boiling of the gravy had so enfeebled its "vegetative force" that it was incapable of generating life. Spallanzani retorted that his long-boiled gravy was perfectly capable of supporting life. However, he believed that the living matter had to be introduced from the outside. He demonstrated this by breaking open one of his flasks and allowing air to enter. In a few days huge numbers of bacteria were present. Needham said that by breaking the flask fresh air was allowed to touch the gravy and that this air brought with it more "vegetative force" which then permitted life to arise from the gravy. No final conclusion to this argument was reached in the eighteenth century.

Pasteur Versus Pouchet

In the nineteenth century F. A. Pouchet described experiments in which he had observed the appearance of bacteria even after considerable boiling, thus adding support for Needham's beliefs.

FIGURE 10-2   Pasteur flask.

Pasteur repeated Pouchet's experiments—often getting *opposite* results, though not always. He then took it upon himself to arrive at a definitive answer to the problem. He designed simpler and simpler experiments. His most famous and conclusive experiment involved the construction of his "Pasteur flask," shown in Figure 10-2. This device gradually convinced the opposition, and the case for spontaneous generation was dropped—until A. I. Oparin and modern biochemists revived it. It is interesting to reflect on the fact that most scientists today believe in the possibility of a form of spontaneous generation, as discussed in Chapter 11.

1. Why do you suppose that people believed that fireflies come from sparks, worms from vinegar or slime, and rats from the mud of the river's edge? Were these just examples of

"old-wives' tales"?_____ Explain._____

_____

2. What is there about science that causes the destruction of such ideas?_____

_____

EXERCISE A

Francesco Redi's experiment concerned the belief (first published by Virgil) that dead meat (dead animals) can give rise spontaneously to the white "worm-like" organisms usually associated with rotting animal flesh. Redi opposed the belief that dead organic matter can give rise to live worms. The first important fact that made him doubt spontaneous generation in this case was the resemblance between the "worms" and fly larvae or maggots. By taking some of the "worms" and keeping them in a closed container, Redi was able to watch them until, after several weeks, they turned into flies. Once he had established the relationship between the "worms" and maggots, Redi was able to perform his conclusive experiment involving the following equipment.

(1)   Beakers.
(2)   Raw meat.
(3)   Gauze.
(4)   Paper.

3. What hypothesis was Redi testing? _____

_____

_____

4. Draw the equipment set up in a manner which you believe would substantiate Redi's hypothesis. Include a control and explain what variables (possible causes) must be controlled.

Drawing

Explanation

You will now have the opportunity to test Redi's hypothesis. We have not mentioned a drawback of the original experiment. As the meat was left to rot, the stench became awful. We will not ask you to sacrifice to the degree that Redi did. At the front table you will find all the ingredients to reproduce Redi's experiment. But instead of meat, you will find a dish of "imitation meat," a molasses and flour mixture. Use this instead of the real meat. Each group of four students will take as much of the following as needed: gauze, paper, beakers, and imitation meat. Use a spoon as a ladle. Set up the experiment on a tray and place your names on a card on the tray. The tray will be placed in a room with an open window. In two or three weeks you will have an opportunity to see what happened.

EXERCISE B

Describe what you think was Pasteur's experiment and draw the setup, using his famous flask. Include (1) hypothesis, (2) experiment, (3) control, and (4) conclusion.

Drawing

Explanation

Your instructor will set up the Pasteur experiment according to suggestions from the class. You will have the opportunity to make observations as to the validity of the experiment in a few days.

11 Development of a Hypothesis I

Origin of Life 2

A hypothesis is a *tentative* statement *suggesting* a relationship between aspects of the environment. It may take the form of a solution to a problem supplying the probable cause of an effect, for example, "I believe that human cancer is caused by a virus." It may, instead, simply tie together certain facts into a potential relationship, for example, "The Democratic candidate will win the next presidential election." (This would be a hypothesis if the prediction were based on such "facts" as the polls indicating the current popularity of the party, the candidate, etc.)

Every hypothesis must have a foundation in fact. Someone who hypothesizes that human cancer is caused by a virus might base his hypothesis on the fact that certain cancers of chickens and mice have been proved to be virus-induced.

Scientists who believe the surface of Mars is hard enough for the landing of a space vehicle base their hypothesis on the known gravity of Mars and on the appearance of Mars' surface as recorded on photographs made by exploratory space vehicles. They would not dare ask an astronaut to land on Mars on the basis of their hypothesis. They would have to send an *experimental* unmanned vehicle to test their hypothesis by landing on the surface.

When you decide to go to a particular movie, you are hypothesizing that you will enjoy it. The facts you base your hypothesis on might be your previous experience with movies on the same subject, your past enjoyment of performances by the same cast or director, and so on.

When the cancer researcher injects a suspected virus into an experimental animal, he is *testing* his hypothesis. If he can prove that those animals with the virus get cancer and those without the virus do not get cancer he is on his way to changing his hypothesis into a fact. *A proven hypothesis becomes a fact.*

When you leave the movies you are in a position to know whether or not your hypothesis was valid. You can state with certainty "I *did* enjoy the movie" or "I *did not* enjoy the movie." Before entering the theater your belief that you would enjoy the movie was tentative; when you left the theater you were *sure* that you either liked or disliked the film.

Many persons carelessly treat hypotheses as facts. This is a dangerous flaw in reasoning, as a hypothesis remains a *tentative* statement until given substance by an experiment. Every hypothesis has an implicit question mark lurking behind it. Only an experiment can remove this question mark.

MODERN CONCEPTS OF SPONTANEOUS GENERATION

The problem of life's origin is fundamental to all rational inquiry into the living processes. It has further pertinence, however, in that it is directly related to systems of behavior and belief which we have inherited from our ancestors who, even in primitive times, found it necessary to conjecture about the origin of life.

Religion is a product of man's need for understanding his position in what is often spoken of as a "cosmic plan," an orderly development of forces in the universe which function according to a pattern as yet unclear to us. All of science is based on the premise that the universe operates according to a pattern of "natural laws." The goal of science is to uncover these laws in order to predict the "behavior" of the universe. Both science and religion accept the premise of order in nature, and both are products of man's groping for a better understanding of his role in relation to an often hostile environment. Science and religion, then, are not necessarily antagonistic; there are scientists who can, and do, believe in God. It is interesting to note that "Western" social pressures and cultural influences have acted to inhibit scientific curiosity about the origin of life, and that most of the early work in this area was done by the Russians, whose antireligious political system probably stimulated such inquiries.

Oparin

A. I. Oparin, in 1923, published a small book entitled *The Origin of Life*. In it he questioned the concept that certain chemical compounds fundamental to life as we know it (organic compounds) were always the products of living things. In effect, he set forth the hypothesis that organic molecules could have existed on the primitive Earth before living organisms were present and, furthermore, that these molecules could have joined together under the influence of ordinary intermolecular forces to form tiny living molecular systems. Thus Oparin brought back an old specter and gave it new life, for he was proposing that living organisms originated by **spontaneous generation**.

1. Can we call Oparin's idea a hypothesis? _____

2. Does the accumulation of "favorable" information necessarily mean that some day

this idea will attain the status of a fact? _____

3. Will we ever know whether or not life arose spontaneously as Oparin suggests?_____

Explain._____

_____

_____

Since Oparin's epoch-making contribution, many scientists have added their conjectures to his, so that an imposing mass of information has been accumulated.

Urey

Harold C. Urey is an American pioneer in the study of the nature of planetary environments, including that of the primitive Earth. One of his major contributions is a suggested reconstruction of the Earth's original atmosphere.

> We do not know the temperature at which the Earth's surface was formed, but it can be stated firmly that the only feasible source of heat for producing a high temperature primitive Earth was its own gravitational energy of accumulation.*

*H. C. Urey, *Aspects of the Origin of Life*, M. Florkin, Editor, Pergamon Press, New York, 1960, p. 8.

We will follow his reasoning in this area. If the Earth accumulated this energy so rapidly that the heat could not be radiated into space, then it could have originated in a molten condition. If the process were slower, the Earth may never have been hotter than a few hundred degrees Celsius.

The data seem to substantiate the latter hypothesis because of the presence in the Earth's atmosphere of such relatively heavy gases as krypton and xenon. At high temperatures, molecules of these inert gases would have had enough kinetic energy imparted to them to reach "escape velocity," that is, the speed necessary to escape the Earth's gravitational pull and fly into outer space. (This reasoning is somewhat weakened by new information which suggests that xenon and krypton may be products of the spontaneous disintegration of radioactive elements and that they are not the original molecules present when the Earth was formed.)

Assuming that the Earth was never much hotter than several hundred degrees Celsius, the following reactions might have occurred somewhere between 1 and 4.5 billion years ago:

(1) Metallic iron and water on the Earth's surface would combine to produce hydrogen gas from the reaction

$$Fe + H_2O \rightarrow FeO + H_2$$

The higher the temperature, the more hydrogen would be produced.

(2) Carbon and hydrogen would form methane (CH_4).
(3) Nitrogen and hydrogen would form ammonia (NH_3).

If the temperature of the Earth was not too high, a reducing (hydrogen-rich) atmosphere might have existed, favoring the production of methane and ammonia. Some of these and similar molecules may have been washed into vast oceans by almost incessant heavy rainfalls which may have lasted for hundreds of centuries. Many of these molecules might have been found in the moisture-laden atmosphere.

No free oxygen could have been available in the Earth's atmosphere for millions of years, since the propensity of oxygen to enter into chemical reactions would have been enhanced during the early, relatively hot centuries of the Earth's formation, and it would have become part of such compounds as water.

Since neither bacteria nor oxygen was present to decompose any newly formed molecules, these molecules may have existed for long periods of time, thus favoring even more complex molecular combinations. However, the forces favoring combinations were counteracted to an unknown degree by the forces favoring the spontaneous disintegration of these same molecules.

Urey paints a picture of a primeval Earth largely covered by storm-tossed oceans of fresh water (Where did the salt of our present oceans come from?) covered by an atmosphere rich in hydrogen, methane, and ammonia. He points to the cold outer planets of the solar system (Jupiter, Saturn, etc.) which have atmospheres probably frozen in their primeval state, consisting in large part of hydrogen, methane, and ammonia.

4. Is Urey's idea about the nature of our primeval atmosphere a hypothesis?_____

Explain._____

_____

5. To what degree does our information about the nature of the atmosphere on Jupiter

and Saturn affect Urey's idea of the Earth's primeval atmosphere? _____

_____

_____

6. Do you believe Urey's idea to be significant in a discussion of the origin of life?

_____ Clearly explain why. _____

_____

_____

Miller

S. L. Miller, a student in Urey's laboratory, was concerned with the implications of Urey's hypothetical atmosphere on the validity of Oparin's suggestion that life originated as the result of the spontaneous aggregation of small molecules and atoms into larger living molecular systems. He attempted to recreate the primeval atmosphere of the Earth by admitting methane and ammonia into a flask containing water vapor and hydrogen. He applied energy to this atmosphere in the form of an electrical spark which simulated an available source of energy in nature, lightning.

His apparatus is shown in Figure 11-1. The water accumulating at point C was drawn off and analyzed. It was found to contain glycine, alanine, and several other amino acids, as well as acetic acid, lactic acid, urea, and other organic compounds.

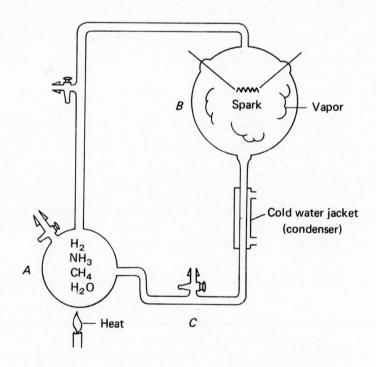

FIGURE 11-1 Experiment with the Earth's primeval atmosphere.

7. This question is designed to test your understanding of the type of reasoning behind Miller's work and your ability to evaluate its significance. On the basis of the preceding discussion, place the appropriate letter beside each of the following statements. If the statement is true, write a T; if it is false, write an F; if it is a hypothesis, write an H; if it is an assumption, write an A.

a. The substances in flask *A* in Figure 11–1 are all organic molecules. _____

b. The gases in the apparatus would combine to form organic molecules. _____

c. Sparks or electricity is necessary for the formation of all organic compounds. ____

d. All organic compounds are made by living organisms. _____

e. It is possible for organic compounds to be formed outside of living organisms. ____

f. The apparatus is solely for the purpose of recreating the original planetary atmosphere and has nothing to do with spontaneous generation. _____

g. Spontaneous generation appears to occur in flask *B* because bacteria are present, are allowed to multiply, and supply the organic molecules. _____

h. Life originated spontaneously. _____

i. In the event that organic molecules are formed in Miller's apparatus, he has proved that life could have arisen on the Earth spontaneously.

j. Life always comes from other life. _____

PRELIMINARY INFORMATION

In order for living molecules to form, certain "building blocks" must be available. These must be able to join with other such molecules to form chains called **polymers**. It is these polymers which are able to perform the complex tasks required of a living system. On this planet all living things are made up of varying combinations of approximately twenty different kinds of these "building blocks" called **amino acids**. All amino acids have the same basic formula: a skeleton consisting of carbon and hydrogen atoms to which are attached one group consisting of a nitrogen atom and two hydrogen atoms (an **amine** group) and another group, called a **carboxyl** group.

Carbon chain (such as might occur when two methane, CH_4, molecules combine)

Amine group

Carboxyl group

The formation of glycine, the simplest amino acid, follows these basic steps:

(1) One of the simplest carbon compounds is a single carbon atom with its four bonds completed with hydrogen atoms:

$$\begin{array}{c} H \\ | \\ H-C-H \\ | \\ H \end{array}$$

Methane, CH_4

Certain replacements of the hydrogen atoms frequently occur in nature. One of these replacements is with the amine group, $-N\begin{array}{c} H \\ \\ H \end{array}$. The result is methylamine,

$$\begin{array}{c} H \\ | \\ H\!\!\diagdown N-C-H \\ H\diagup \quad | \\ H \end{array}$$

(2) The *opposite* hydrogen is replaced with a carboxyl group, $-C\begin{array}{c} O \\ \\ OH \end{array}$, making the compound an organic acid. Since the compound contains an amine group and is an acid, it is called an amino acid:

Amine group $\begin{array}{c} H \\ | \\ H\!\diagdown N-C-C \diagup\!\!^O \\ H\diagup \quad | \quad \diagdown OH \\ H \end{array}$ Carboxyl group

Glycine

All other amino acids follow the basic pattern of glycine. The upper H, however, is replaced with a variety of carbon chains, giving each of the different amino acids its distinct characteristics. The general formula for all amino acids except glycine is

$$\begin{array}{c} R \\ | \\ H\!\diagdown N-C-C\diagup\!\!^O \\ H\diagup \quad | \quad \diagdown OH \\ H \end{array}$$

where R represents a carbon chain.

The replacement of the upper H with different kinds of carbon chains is the reason for the variety of amino acids. For example, a three-carbon chain would result in a different kind of amino acid than a four-carbon chain replacement.

Notice that an almost infinite variety of amino acids is possible, each amino acid varying from the other by the addition of another carbon, sulfur, or phosphorus atom, or even by changing the geometric configuration of the carbon chain. On our planet, however, approximately twenty different kinds of amino acids have been found in protoplasm; all protoplasm is composed of some combination of these.

Amino acids may be made to join together into chains of varying lengths and characteristics. The attachment always occurs in the same fashion. Enough energy is applied to force out a molecule of water from the two molecules, the amine contributing an **H** and the carboxyl contributing an OH (yielding HOH, written H_2O), as shown below:

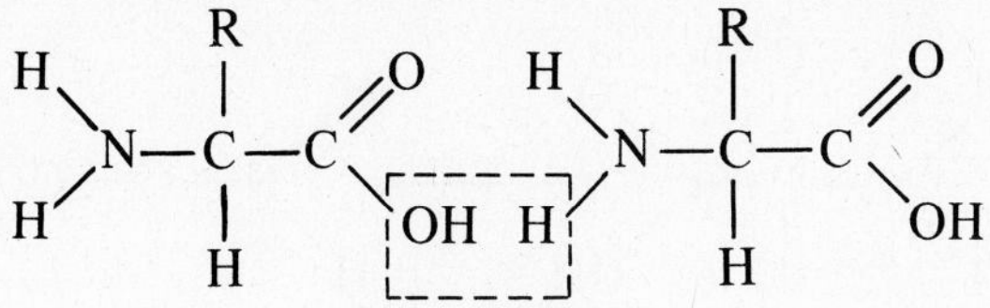

Water molecule

When the water molecule is removed, the carboxyl carbon and the amine nitrogen each have an extra bond available (the

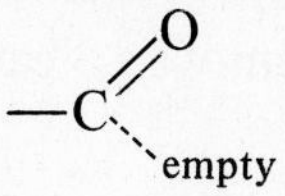

has only three of its four available carbon bonds filled; and the

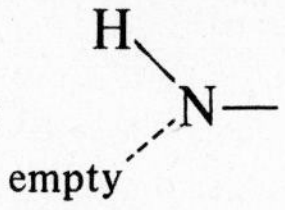

has only two of its three bonds completed). The carbon of one amino acid then joins to the nitrogen of the other amino acid in what is called a peptide bond:

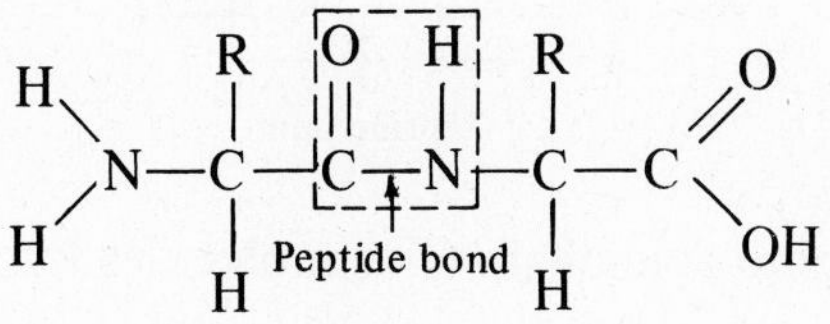

Peptide bond

When many amino acids are joined in a chain (a polymer), the resulting long molecule is called a **protein**.

EXERCISE

Assume that you have available an abundance of each of the specified molecules. Join them together into the required chains in the space provided.

   Given: Five-carbon molecules (C_5H_{12})
   Three-carbon molecules (C_3H_8)

Amine groups

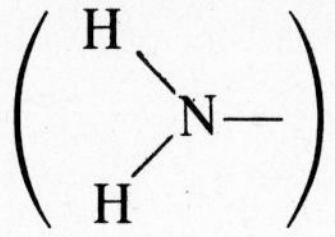

Carboxyl groups

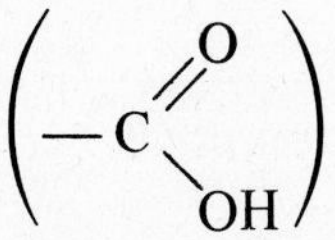

Use at least one of each type of molecule.

Required: An amino acid chain consisting of five amino acids joined together by peptide bonds. Use two five-carbon amino acids and three three-carbon amino acids. You should end up with five amino acids joined together, following the steps shown below:

Step 1: Formation of a water molecule from the COOH of one amino acid and the NH₂ of another.*

Step 2: Formation of peptide bond. The box shows the region where two amino acids are joined. Note that the water molecule has been removed, leaving $-\overset{O}{\underset{||}{C}}-$ and $-\overset{H}{\underset{|}{N}}-$ joined together.

Draw your completed amino acid chain (five molecules long).†

----

*Each line around a C represents a bond to a hydrogen atom; that is,

$$-\overset{|}{\underset{|}{C}}- \ = \ H-\overset{H}{\underset{H}{C}}-H$$

†It will be easier for you to draw your amino acid chain if you first draw the whole chain with each water molecule still attached, then you can draw your finished molecule above. Use scrap paper for your first drawing.

Draw a box around each peptide bond in your completed polymer. If models or styrofoam balls are available, put together your polymer so as to get a three-dimensional understanding of molecular structure.

8. S. L. Miller obtained some amino acids in his solution. Discuss clearly your opinion as to whether or not spontaneous generation occurred. _____

_____

_____

_____

9. What *can* Miller say about spontaneous generation at the conclusion of his experiment, since he cannot claim that he has shown conclusively the manner in which life originated? _____

_____

_____

10. If you had a laboratory at your disposal, what would be your next steps in the hope of arriving at an understanding of the origin of life? Suggest a hypothesis and an experiment designed to test your hypothesis.

V Vertebrate Anatomy

12 "Mental Discipline" in Science

Skeletal System

In the nineteenth and early twentieth centuries, one of the guiding principles of education, especially in science, was the belief that continued practice in the use of one or another of the mental faculties would result in a strengthening or "disciplining" of that faculty. Furthermore, it was believed that the "disciplined" faculty would be more effective when applied to "life" (that is, nonschool) situations. For example, the study of mathematics was supposed to endow students with a type of logic which would be useful in any situation requiring logical analysis.

Scientific training consisted largely of the rote memorization of myriads of facts. A fringe benefit of this accumulation of knowledge was the supposed enhancement of the student's memory just as the effectiveness of a muscle is increased by using it.

In this context it may be considered unfortunate that the brain is not a muscle and that the accumulation of knowledge is not necessarily an aid to memory. There have been repeated attempts to validate the hypothesis that there is a transfer of training from the school situation to everyday life. While it has not been proved that the brain is made more effective by specific attempts to discipline it, modern researchers have been uncovering clues which suggest that, to a limited degree, transfer of training does take place. For example, in the following exercise you will be asked to memorize the names and positions of all the bones in the body. There are two benefits to be derived from this type of task, and to some degree these may be transferable. They are

(1) You will find that what seems to be an insurmountable task—a feat of prodigious memory—is easily accomplished. You will have a new understanding of your own mental capacities and will approach other problems requiring memorization with more confidence.

(2) You will develop a technique for memorizing related information. You will have the opportunity to try out different ways of organizing your information to see if one way is more effective than another. Should you try to memorize the bones alphabetically? Starting from the head and working down? Or by sections of the skeleton?

The techniques which you develop will perhaps be applicable to the memorization of your verb conjugations in French class, or names and dates in History. (Let us hope so.)

"Transfer of training" and "mental discipline," as techniques for developing your mental faculties by strengthening portions of your brain, have been shown to be invalid hypotheses. On the other hand, some improvement of your generalized learning abilities can accrue from practice. It is likely to increase your self-confidence and to help you learn techniques for solving problems.

PRELIMINARY INFORMATION

The skeleton in vertebrate animals serves as an internal supporting mechanism to which are attached the voluntary muscles. Movement is accomplished by the contraction of these muscles. Another function of the skeletal system is the protection of the softer respiratory, digestive, and nervous organs. The skull is composed of fused flat bones which effectively protect the brain from damage. The spinal column is made up of relatively massive bony structures called **vertebrae**. Each vertebra has a small hole in it, through which runs a portion of the spinal cord. The skull and spinal column thus protect much of the nervous system, which might otherwise be easily damaged.

Some animals have neither backbones nor internal skeletal systems. These are the **invertebrates**. Since they do not have bones for the muscles to pull on, how do they move? Be prepared to discuss the mode of movement of molluscs (clams and squid), arthropods (insects and lobsters), and cnidarians (jellyfish).

The skeletal system of man may be divided into two basic portions—the axial skeleton and the appendicular skeleton. The groups of bones associated with each are

> **Axial skeleton.**   Skull, vertebrae, ribs, sternum
> **Appendicular skeleton.**   Pectoral and pelvic girdles and the appendages attached to them.

Much of the mobility and flexibility of the body is made possible by the following types of skeletal joints.

*1. *Immovable,* such as the sutures of the skull. Why is it necessary that these joints be

movable at some time during the life of the organisms? _____

2. *Slightly movable,* as found in the pubic symphisis and vertebrae. Why must each of

these be slightly movable? _____

3. *Movable* such as the *hinged joints* of the elbow, the *ball and socket* joints of the hip and shoulder, and the *gliding joints* found in the ankles and wrists. Why must these joints be

highly movable? _____

EXERCISE A

Below will be found the names of the bones of the body. You will be held responsible for learning to recognize each bone and describing where the bone is found. You will not be permitted to use any reference materials while in the laboratory. You may, however, do whatever studying, sketching, or note-taking you wish during the interval between this laboratory period and the next. You may make notes if you wish on the pages of this laboratory manual, but do not bring with you any published reference material. At the conclusion of the next period you will be given a practical quiz on the skeletal system. No references (including this manual) are to be used when you take the test.

Each group of students will receive a box of human bones. Lay these out on your table top, using the articulated skeleton in the room as a frame of reference. Quickly sketch as many bones as you have time for, making sure to indicate the characteristics which differentiate each from the others.

---

*Come back to these questions after you have studied the bones if you cannot answer them now.

Name of Bone	Sketch	Notes

SKULL

1. Frontal

2. Parietal

3. Temporal

4. Zygomatic

5. Maxilla

6. Mandible

7. Sphenoid

8. Nasal

9. Occipital condyles

10. Foramen magnum

VERTEBRAE

11. Atlas

12. Axis

Name of Bone	Sketch	Notes

13. Cervical vertebrae

14. Thoracic vertebrae

15. Lumbar vertebrae

16. Sacral vertebrae

17. Coccygeal vertebrae

PARTS OF VERTEBRAE

18. Neural spine

19. Centrum

20. Anterior and posterior
    zygopophysis
    (Superior and inferior
    articular processes)

21. Neural canal

22. Transverse processes

GIRDLES

23. Scapula

24. Clavicle

Name of Bone	Sketch	Notes

25. Acromion (process)

26. Ilium

27. Ischium

28. Pubis

29. Pubic symphysis

APPENDAGES

30. Patella

31. Humerus

32. Femur

33. Radius

34. Tibia

35. Calcaneus

36. Metacarpals

Name of Bone	Sketch	Notes

37. Phalanges

38. Carpals

39. Ulna

40. Fibula

41. Tarsals

42. Metatarsals

OTHER BONES

43. Ribs

44. Sternum

Ossicles of ear
(available in a plastic
block at front table)

45. Malleus

46. Incus

47. Stapes

EXERCISE B. FUNCTIONS OF BONES

Each bone, as you have seen, differs in shape from the others. Can the appearance of a bone be related to its function? Examine the bones and answer the following questions about their function.

4. What is the significance of the round head of the radius? _____

_____

_____

5. Discuss the implications of the fact that man and the other primates are the only animals that have thumbs which can move at right angles to the rest of their fingers (opposable thumb). What are the evolutionary, anthropological, and sociological implications of this?

_____

_____

_____

6. Look at the teeth in a human skull, the skull of a rat, and the skull of a dog.

a. Are all the teeth in the human skull the same?_____Sketch each kind of tooth and relate its shape to its function—for example, how is a tooth adapted for the chewing of one kind of food rather than another?

     b. Do the rat and dog have the same kind of teeth as the human? _____

Explain. _____

_____

_____

_____

    7. Note the size and shape of the chamber in which the brain is found in the rat, dog, and human skulls. Discuss the implications of this, especially the relationship of the *shape* of the skull to the intellectual capacity of the brain inside it. _____

_____

_____

_____

    8. Compare the positions of the eye sockets in the human and rat skulls. Describe the difference in their respective positions and the effect of each on *vision*. _____

_____

_____

_____

    9. Discuss the nature and meaning of the coccyx. _____

_____

_____

_____

    10. a. Discuss the nature and importance of the curvature of the spinal column. _____

_____

_____

     b. Why is the spinal column composed of many small bones instead of one long one?

_____

_____

c. Discuss the advantages and disadvantages of erect posture. Consider, for example, the different positions of the internal organs in a two-legged animal and in a four-legged animal.

_____

_____

11. Relate the attachment of the ribs to their function. _____

_____

_____

_____

12. Explain the interaction among the atlas, axis, and occipital condyles. _____

_____

_____

_____

13. Man runs on the soles of his feet, dogs and cats run on their toes, and hoofed animals run on the tips of one or two modified toes. Hoofed animals also have elongated legs, and their wrists and ankles are raised from the ground so that the ankle corresponds in position to our knee and the hoof is the toenail of one or two toes. Discuss the functional implications of each of the above evolutionary adaptations to movement.

13 Organization and Application of Information

Digestive and Respiratory Systems

One of the few correct aspects of the scientist's image projected to the general public is that he is an extremely well-organized person. Like all stereotypes (generalizations based on insufficient data into which all observations of a particular class are *forced* to fit), this crude picture has its weaknesses. Scientists outside the laboratory (as private citizens) are just as prone as their neighbors to be unorganized in their daily lives. But in a scientific situation, the rigors of efficient thinking force the scientist to arrange his thoughts and data as effectively as possible.

This exercise requires that you learn to identify the digestive and respiratory organs of the fetal pig and man. You will have only one period in which to accomplish this. You will be placed in a situation which will tax your abilities to complete your assignment, forcing you to search for the most effective means of learning. It is hoped that the information *about* the learning process, and the learning techniques you develop, will be useful to you in other situations requiring efficiency and economy of thought.

The key to efficient performance of this exercise is in the organization of the material. The benefit of placing your information in a logical sequence is demonstrated by the number systems below.

System A	System B
1	10
13	20
9	30
7	40
24	50
6	60
15	70
8	80
26	90

2. Memorize each list, recording the amount of time it took to reproduce each perfectly.

Which sequence is easier to remember? _____ Why? _____

_____

_____

Increased efficiency is not the only benefit of meaningful, systematic organization. Learning is also enhanced by the logical arrangement of facts. Attempt to memorize the word sequences below.

3. Record the number of seconds it takes to learn each sequence *completely* with no mistakes.

System C	System D'	System E
art	zeb	Three
friend	rin	cheers
song	tup	for
then	gik	the
boy	ork	red
elbow	fut	white
fish	zilch	and
toy	slarg	blue

No. of
seconds: _____   _____   _____

4. Which system is easiest to remember? _____   Which is most difficult? _____

Why? _____

_____

_____

PRELIMINARY INFORMATION

The digestive system includes all the organs whose functions are related to the breaking down of relatively large molecules called "food." Associated with the actual breakdown process are the other major functions of the digestive system: (a) the secretion of enzymes which accelerate the disintegration of food molecules and (b) the absorption of the products of digestion. (The small molecules resulting from the digestion of food pass through the walls of the small intestines into the bloodstream.)

Essentially, the digestive tract (alimentary canal) is a tube which begins at the mouth and ends at the anus. The mouth, stomach, small intestines, pancreas, and liver pour their secretions into the alimentary canal and thus facilitate digestion.

The function of the respiratory system is to bring about an exchange of the gases associated with the oxidation or burning of food molecules. The major source of energy (fuel) of the body is the glucose molecule, $C_6H_{12}O_6$. This simple sugar is oxidized in the cells of the body to produce the energy necessary for the performance of the life functions.

The reaction produces two waste gases which, if allowed to accumulate, would disrupt the very precise balance of substances in the body and cause death to occur.

The equation for the oxidation of glucose is

$$\underset{\text{Glucose}}{C_6H_{12}O_6} + 6\,O_2 \longrightarrow energy + \underset{\text{Waste gases}}{6\,CO_2 + 6\,H_2O}$$

Our respiratory system provides, in the lungs, a moist membrane through which the molecules of each gas move. If there is a large concentration of carbon dioxide and water in the blood surrounding the moist membrane, many molecules of these gases will pass from the blood through the membrane into the air in the lungs, to be expelled as we exhale. At the same time, oxygen passes through the membrane into the blood, to be carried to the cells.

The other respiratory organs (diaphragm, trachea, etc.) are concerned with the mechanical process called **breathing**, a method used by certain vertebrates to draw in and expel air. All animals and plants must carry on respiration *all the time* in order to obtain the energy required for survival. Some use lungs and breathe, others use gills, moist skins, or, in plants, tiny holes in the leaves called **stomates**.

The Fetal Pig

In this laboratory exercise you will investigate the arrangement of organs in the abdominal and thoracic (chest) cavities of the fetal pig. We use this animal because we have a ready supply of unborn pigs obtained from slaughterhouses and injected with a liquid rubber compound which defines the arteries and veins. Since there is no market for unborn pigs, except as laboratory animals, it is possible for us to obtain large numbers of these fine specimens (they are much closer in their anatomy to the human than are rats or frogs) at comparatively low cost.

1. Hypothesize the reasons for the absence of a commercial food market for fetal pigs. Do you think they are poisonous? Do you think their taste is unpleasant? Explain the

reasoning behind your hypothesis. _____

_____

_____

_____

_____

_____

While these specimens have not yet been born, they have developed to a stage where they would be able to function if they were born, and their organs are clearly enough defined for study.

The animals are preserved in alcohol or formalin and are leathery to the touch. They have not been sacrificed for use in the laboratory, since they would have been discarded by the meat packers if we had no use for them.

EXERCISE

The mammalian digestive and respiratory organs are listed on the following page together with some remarks about their appearance and functions. After studying this list, you are to describe the manner in which you will organize the materials so that you will be able to learn it most effectively. You will then examine a fetal pig, as directed on pages 153-54.

The Respiratory and Digestive Organs of a Mammal

Tongue: Muscular organ in the mouth which tastes food and helps prepare it for swallowing.

Liver: Large, reddish brown, lobed organ below the diaphragm.

Pharynx: The "throat" or chamber at the rear of the mouth common to the respiratory and digestive systems.

Glottis: A thin slit at the base of the pharynx which is the opening of the trachea. A respiratory structure.

Lungs: Large, flesh-colored, spongy respiratory organs. Most obvious organs in the chest cavity. Paired.

Duodenum: First section of the small intestines. The portion which originates at the stomach.

Stomach: Sac-like, flesh-colored organ just below the diaphragm. The esophagus enters the upper end and the duodenum leaves the lower end.

Ileum: The portion of the small intestines adjoining the large intestines.

Trachea: A shiny white tube which carries air to the bronchi and thence to the lungs. In the neck and upper chest.

Soft palate: The posterior portion of the roof of the mouth. Soft to the touch.

Rectum: The most posterior portion of the large intestines. Has muscular walls and opens to the outside through a valve called the anus.

Diaphragm: A sheet of muscle separating the chest cavity from the abdominal cavity. Just above the liver and stomach and below the lungs. Controls the volume of the chest cavity (with rib cage) and thus makes breathing possible.

Bronchi: Two forks from the trachea carrying air to the lungs.

Gall bladder: A greenish sac on the underside of the liver. Has a duct which carries bile into the duodenum.

Larynx: The voice-box or adam's apple. A swelling on the trachea at its upper end.

Region where **ileocolic valve** might be: a valve at the junction between the small and large intestines. Not visible. Regulates the flow of materials from the small intestines to the large intestines.

Esophagus: A flesh-colored tube behind (above) the trachea which leads from the pharynx to the stomach.

Cecum: A blind pouch at the beginning of the large intestines. Part of the large intestines which begins at the region of the ileocolic valve. (Search carefully for this.)

Pancreas: A whitish organ which appears to be composed of many tiny lobes. Just under the stomach in a **C**-shaped space bounded on the top by the stomach and underneath by the duodenum. Secretes enzymes for digestion and the hormone insulin.

Hard palate: The hard front portion of the roof of the mouth.

Region where **appendix** might be (if pigs had one): On the cecum. Check the cecum of the manikin for a small finger-like projection.

Large intestines: Greyish, wrinkled, and thicker than the small intestines. In the human they form a ∩ shape. In the pig they are a compact mass of grey tubes.

Pyloric valve: Regulates the flow of material from the stomach to the duodenum. Cannot be seen, but can be felt as a hard, muscular region at the junction of the stomach and the duodenum.

Epiglottis: A rubbery structure at the base of the tongue that covers the glottis during swallowing, preventing food from entering trachea.

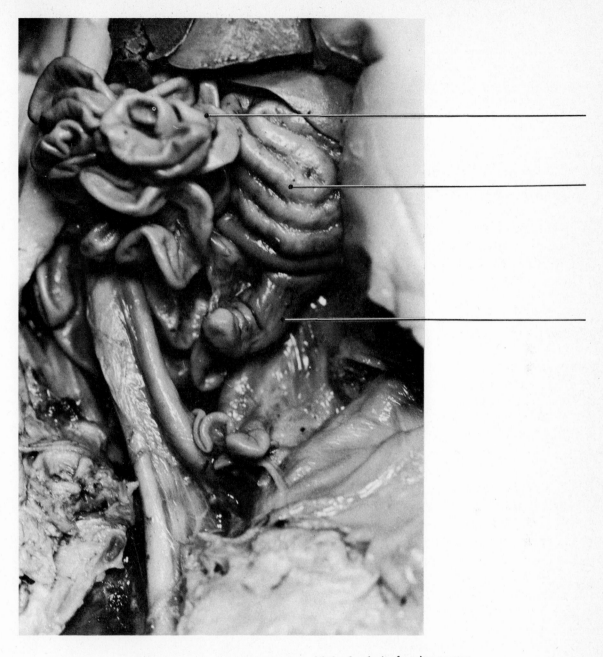

FIGURE 13-1   Abdominal organs of the fetal pig showing cecum.

Directions

Place the pig in the dissecting pan with the underside (belly) facing you. With a slipknot attach a double-length rubber band to a foreleg and slip it under the pan. Loop the free end over the other foreleg. Now repeat this process with the hind legs. Spread the legs out so that the internal organs are clearly visible.

Since the pigs have already been predissected for you, it will not be necessary to use any instruments other than a wooden or metal probe. Use the probe to carefully push aside any organs which prevent a clear examination of the alimentary and respiratory organs.

Work with your partner. Make sure you can identify all of the organs on your list in both the fetal pig and the human manikin. When you have finished your dissection, label the lines on Figures 13–1 and 13–2.

The last half hour of the period will be devoted to a practical examination of your success at finding each organ. Your instructor will call you to the front of the room. Do not bring your pig or written material with you.

Plan your organization of digestive and respiratory organs below.

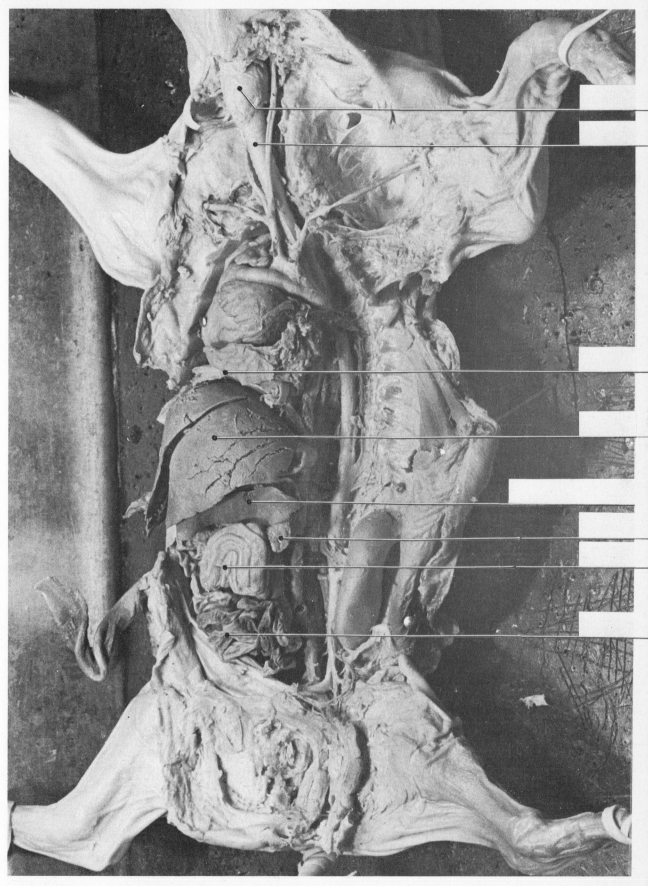

FIGURE 13-2 Thoracic and abdominal cavities of the fetal pig (stomach and lungs not visible).

14 Requirement of Accuracy in Science

Circulatory System

Scientific discoveries are products of the cooperation of scientists around the world. At any one time, facets of a problem may be under simultaneous investigation in Stockholm, Tokyo, London, and Los Angeles.

In order to accept as facts the statements of other investigators, it is often necessary to repeat their work, thus fulfilling the requirement of science that all statements of fact must be empirically verifiable. All published reports of scientific work must include directions for the repetition of the experiment. Often, scientists repeating the experiment obtain different, even contradictory, results. They publish contradictions of the data originally reported, thus throwing out a challenge for a scientific controversy with the original investigator. The battle rages in the form of reports, charts, and tables, revealing little of the true drama or effort expended by the two antagonists. Finally one group is proven incorrect. Their undoing is sometimes due to inaccuracies resulting from inadequate attention to detail in their experiments. For example, when Needham and Spallanzani had their controversy concerning the spontaneous generation of bacteria, both used sealed flasks. Needham sealed his flasks with a more or less porous cork, while Spallanzani melted his flask shut. Needham often found life (bacteria) in his flasks; Spallanzani rarely found life in his. When Spallanzani repeated Needham's experiments he boiled the contents of his flasks for a long time and used an effective closure on his containers, which Needham had failed to do. He repeated Needham's experiments almost exactly. The differences seemed inconsequential to scientists of the period who had no knowledge of bacterial spore formation or the presence of bacteria on dust particles. It was, however, this small variation between the methods of the two experimenters that resulted in the questioning of the then-acceptable premise that bacteria originated spontaneously.

It is not often that the failure to observe details with precision affects our thinking so significantly. More often, in our daily lives, variations due to imprecision may bring about more commonplace disturbances such as burned roasts, ill-fitting clothing, or even lamentable political decisions.

One way for you to develop the habit of precise attention to detail is to practice. This laboratory exercise is designed to give you this needed practice.

PRELIMINARY INFORMATION

Many of the most profound scientific thinkers of antiquity applied themselves to the problem of determining the mechanism of blood circulation. Aristotle and one of his most influential interpreters, Galen, believed that at least part of the substance which flows through the blood vessels is similar to air and contains a vital force which sustains life. Galen named this mysterious material **pneuma** and suggested that it infiltrated through the wall separating the right and left ventricles of the heart through invisible pores. This concept was perpetuated through the Dark Ages because of the unquestioning acceptance of Galen's interpretation of circulation as a kind of scientific dogma which was to be believed without question.

Even in the early Renaissance, when ancient concepts were vigorously questioned, the Galenic interpretation of circulation was accepted. One reason for this was the fact that dissection was considered vulgar by most physicians and biologists, and was left to barbers to perform while a lecturer sat high above the operating table giving directions. It was because of his disgust with the slovenly way in which a barber was following his directions that Paracelsus, leaping off his podium in a rage and seizing the scalpel himself, set the stage for the true interpretation of the circulatory system. His careful attention to the accurate rendition of directions for dissection made impossible the perpetuation of the vague interpretations of the ancients.

Finally, William Harvey, in 1628, suggested that the path of the blood was circular—that it flowed to the lungs, where it obtained oxygen and released wastes, and then back to the heart, from which it was pumped to the rest of the body, returning to the heart through the veins. The fact that there are valves in the veins which prevent the blood from flowing away from the heart led Harvey to conclude that the veins always carry the blood *to* the heart. He also deduced from this premise that the arteries carry the blood *away* from the heart.

A weakness in Harvey's argument was the problem of the connection between the two types of vessels. No one had yet discovered capillaries. It was not long before these vessels were observed, however, since there was much evidence pointing to their existence.

■ **THE PROBLEM:** *To learn the positions and functions of the blood vessels of the human and the fetal pig.*

□ Subproblem 1: *To understand clearly and to be able to draw the path of blood in the pulmonary circulation, including the appropriate chambers of the heart.*

□ Subproblem 2: *To understand the special function of the hepatic portal system and to be able to describe its vessels.*

□ Subproblem 3: *To understand and to be able to trace the fetal modifications of the circulatory system, including the* foramen ovale, *the* ductus arteriosus, *and the* ductus venosus.

Note: At the end of this exercise you will be held responsible for

(1) Tracing a drop of blood throughout the body, given only its point of origin and its destination.
(2) Identifying all the blood vessels mentioned below, either in the fetal pig or in the human manikin.
(3) Drawing a *schematic* representation of the blood vessels of the body.

You are not to consult any textbook in the laboratory, although you may bring notes you have taken from books to help you over a few knotty problems.

EXERCISE

Find the blood vessels listed on the following pages in the fetal pig and in the human manikin. Then label the lines on Figures 14–1 through 14–7. You may also use the photographs to help clarify the positions of the vessels.

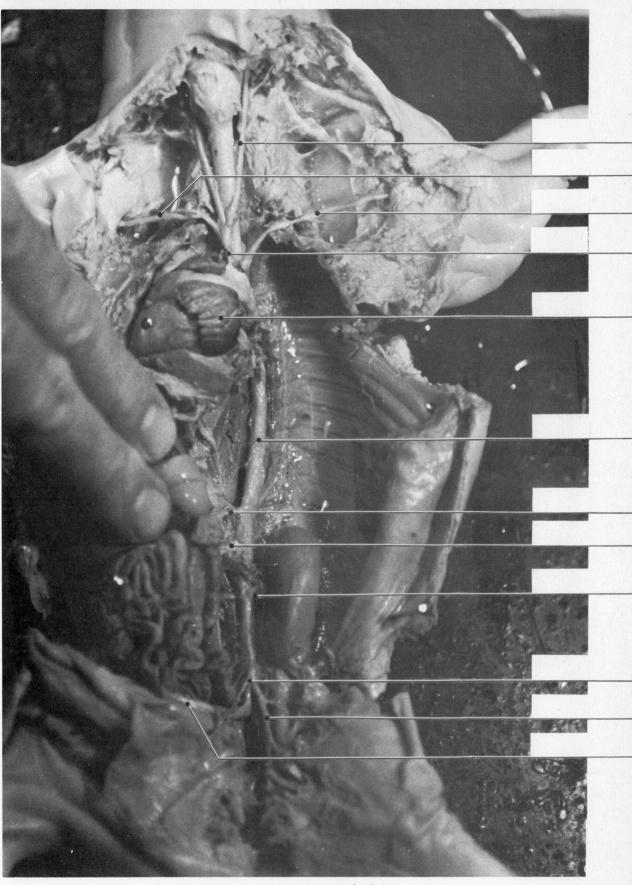

FIGURE 14-1 Arteries of the fetal pig.

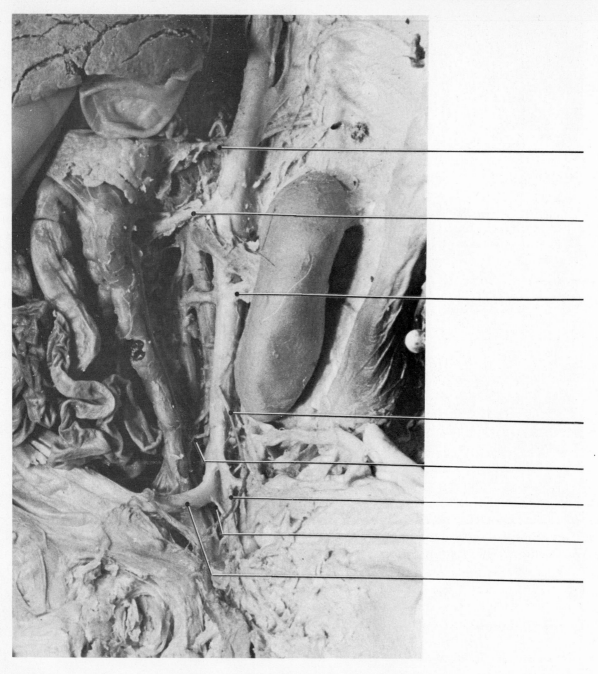

FIGURE 14-2 Abdominal arteries of the fetal pig.

Arteries

Your instructor will show you the dorsal aorta in the pig to provide you with a point of departure.

Dorsal aorta: Progresses from the left ventricle in an arch over the heart and down the dorsal wall of the thoracic cavity.

Aortic arch: The aorta arches as it leaves the left ventricle. In the pig it has two vessels branching anteriorly. In man there are three vessels.

Pulmonary artery: From right ventricle. Clearest vessel coming from heart. Crosses to join aorta (ductus arteriosus) and *down* to lungs.

Brachiocephalic artery: First branch of aortic arch.

FIGURE 14-3 Arteries of the lower abdomen of the fetal pig.

Right subclavian artery: From brachiocephalic into right shoulder.

Left subclavian artery: Second branch off aortic arch. Goes to left shoulder.

Right and left common carotid arteries: Right and left forks of brachiocephalic just below neck.

Right and left internal carotid arteries: Divide from common carotids in region of jaw. Innermost carotid vessels. (Include these in your tracing question, but do not try to find them in the pig unless you have time to spare at the end of the period.)

Right and left external carotid arteries: Divide from common carotids in region of jaw. Branches closest to outside of body. (Include these in your tracing question, but do not try to find them in the pig unless you have time to spare at the end of the period.)

Coeliac artery: Branches from the aorta toward the right. First branch beneath the diaphragm. Breaks into hepatic, splenic, and gastric arteries.

Anterior mesenteric artery: Second branch off aorta below diaphragm. To anterior portion of small intestines.

Right and left renal arteries: Branches from aorta to kidneys.

Right and left genital (ovarian or spermatic) arteries: Slender vessels in lower abdomen. Last *paired* vessels coming from aorta.

Posterior mesenteric artery: Last small, single vessel off aorta. In lower abdomen.

Right and left external iliac arteries: Large vessels branching off aorta in region of rectum, just before final fork of aorta. Penetrate upper thighs and extend down leg.

Right and left internal iliac arteries: Final fork of aorta, in region of rectum. Become umbilical arteries a short distance from junction with aorta. Then continue down leg as slender vessels coming off umbilical arteries as they turn at right angles to leave the body in the umbilical cord.

Right and left umbilical arteries: Progress from right and left internal iliacs in upper thigh, along ventral body wall, and into umbilical cord.

Coronary artery: Small vessel running diagonally across center (coronary sulcus) of ventral surface of heart.

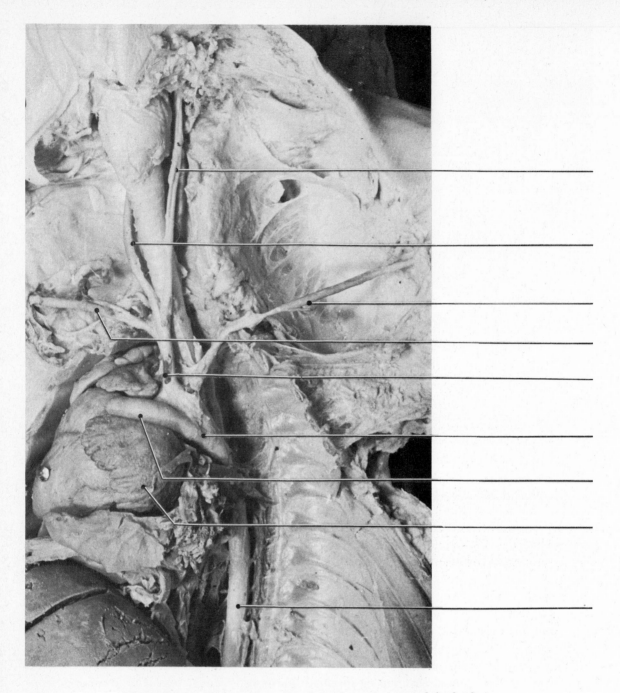

FIGURE 14-4   Heart and arteries of the thoracic cavity of the fetal pig.

Veins

Posterior vena cava: Runs alongside abdominal aorta next to liver, and enters lower right atrium.

Anterior vena cava: Large vessel which enters top of right atrium. Formed by fusion of right and left common jugular veins.

Right and left common jugular veins: Short, wide vessels which form the arms of a Y. Fuse to form anterior vena cava.

Right and left external jugular veins: Drain face. Run down neck to common jugulars. Outermost of the two kinds of jugular veins.

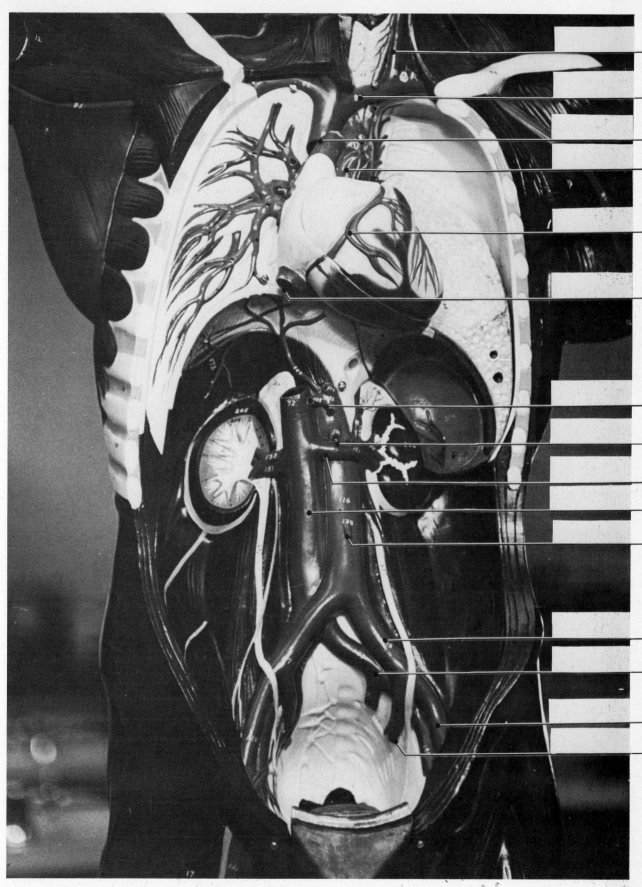

FIGURE 14–5 Human arteries and veins.

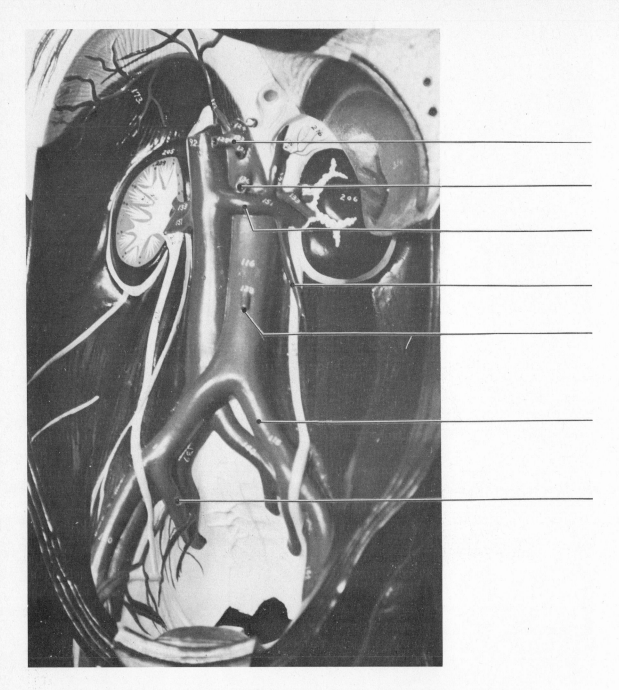

FIGURE 14-6 Human abdominal arteries and veins (genital arteries not shown).

Right and left internal jugular veins: Drain brain. Run along trachea to join common jugulars near heart.

Right and left subclavian veins: From shoulders and arms to common jugular. At right angles to jugulars.

Hepatic vein: From liver to posterior vena cava (not clearly visible).

Hepatic portal vein: Fusion of veins from digestive system (anterior mesenteric, gastric, splenic, and others) to enter liver. Not usually visible, except through careful dissection. (Do not search for this unless you have time at the end of the period.)

Right and left renal veins: From kidneys to posterior vena cava.

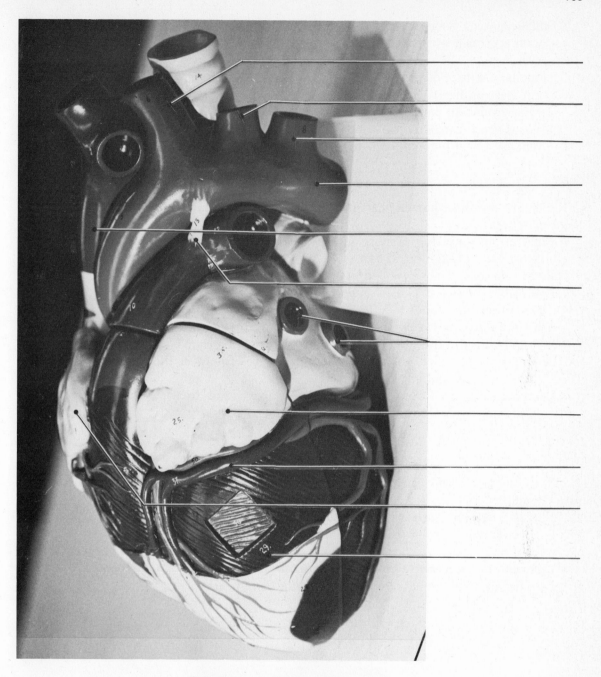

FIGURE 14-7   Model of human heart.

Right and left genital (ovarian or spermatic) veins: From ovaries or testes to posterior vena cava.*

Right and left common iliac veins: In region of rectum; make fork at base of posterior vena cava.

Right and left internal iliac veins: At junction of thigh with body. Join common iliacs; fork toward midline of body.

Right and left external iliac veins: Large vessels joining common iliacs near internal iliacs; fork toward outside of legs.

*The left ovarian or spermatic vein often leads to the left renal vein in pigs and certain other mammals.

Umbilical vein: Single vessel running through umbilical cord to liver (probably cut in your specimen when abdomen was opened). Look for cut ends at umbilical cord and liver.

Pulmonary vein: Four small veins from lungs to left atrium. Not clearly visible in pigs, but clear on heart models.

Coronary vein: In coronary sulcus under coronary artery.

Azygous vein: Drains dorsal wall of thoracic cavity. Runs from dorsal wall to anterior vena cava near junction with right atrium.

Heart

Examine model of human heart in manikin for maximum detail. See also demonstration of sheep or beef heart.

Right atrium (auricle): Grayish sac anterior to right ventricle. Seems to be attached to heart and not an integral part of it.

Right ventricle: Thinner-walled of the two ventricles. Below right atrium.

Left atrium (auricle): Grayish sac on anterior of heart on left side (pig's left; your right).

Left ventricle: Posterior to left atrium. Largest thick-walled chamber.

Tricuspid valve: Between right atrium and right ventricle.

Bicuspid (mitral) valve: Between left ventricle and left atrium.

Semilunar valves: Sets of half-moon-shaped valves found in the entrances of the pulmonary artery and aorta.

Chordae tendinae: Cords preventing the bicuspid and tricuspid valves from turning inside out (back into the atria).

Fetal Modifications

Your instructor will briefly describe the mechanics of these variations from the post-birth circulation.

Ductus arteriosus: Between pulmonary artery and aorta.

Ductus venosus: Extension of umbilical vein through liver to hepatic vein and posterior vena cava.

Ligamentum arteriosus: Remains of ductus arteriosus in adult. No function.

Foramen ovale: Hole between left and right atria of heart.

15 Accurate Description and Communication in Science

Urogenital System

In this era of jet travel, the weakening of geographical boundaries has underscored the conflicting philosophies and ideologies which developed over thousands of years when nations were isolated from one another. Now, suddenly thrown together by political and economic necessity, as well as by airplanes, nations face the increasing problem of adequate communication. Wars have been fought over "misunderstandings."

The scientist is faced with problems of communication which are aggravated by the fact that those who deal with a common body of facts must have more or less identical concepts of the nature of those facts. When scientists deal with uranium, for example, they must very carefully describe which isotope they are studying because U-235 is fissionable and can be used for atomic reactions, while U-238 is not fissionable and therefore cannot be used for the same end.

The social scientist has problems that go beyond those of his colleagues in the natural sciences whose facts are usually more readily observable and can be measured with rulers or scales. The social scientist must try to describe and measure such concepts as "prejudice," "authoritarianism," "psychosis," etc. If he cannot describe these phenomena in a manner permitting others to have a common understanding with him, the social scientist cannot function, for he would be violating one of the cardinal tenets of scientific endeavor; he would be representing something as a "fact" which could not be corroborated by others.

To overcome this difficulty in part, the social scientist uses the **operational definition**. By "operationally defining" his "facts" he assures himself that others dealing with the same phenomena see them as he does, at least for the purposes of exposition. An operational definition, then, is a statement of the manner in which the scientist views a phenomenon in a functional sense—that is to say, in a manner which allows him to "operate" or function scientifically. For example, a sociologist might define poverty in America as "a level of subsistence based on an income of less than $9700 for a family of four." While this definition is not all inclusive, it is useful because it clearly defines the concept "poverty" in terms which can be understood without ambiguity.

1. Write an operational definition of the term "juvenile delinquent." _____

_____

_____

_____

2. Your instructor will write on the chalkboard a few of the definitions of the term

"juvenile delinquent" contributed by members of the class. Are they all identical?_____

Are some of them remote in meaning from one another?_____If members of the class were to discuss "juvenile delinquency" without operationally defining the term before they

began their discussion, would they all be talking about precisely the same concept?_____

Accurate Description in Biology

In today's laboratory, it will be your responsibility to describe the male and female urogenital systems of the fetal pig. Errors in description caused by a slight misinterpretation of the directions or by carelessness will be compounded because subsequent steps are based on the correct interpretation of previous directions. For example, in your study of the circulatory system, you will recall that incorrect identification of the aorta as the pulmonary artery would have rendered impossible the proper determination of the branches of the aorta. Your purpose is to describe each organ so clearly that it would be impossible to mistake one for another.

3. What characteristics can you use to enhance the quality of your descriptions? What characteristics of objects are suitable for use in differentiating them from other objects?

_____

_____

_____

_____

_____

■ **THE PROBLEM:** *To describe the paths of the sperm and egg from their points of origin to the outside of the body.*

Your description should fulfill the following criteria.

(1) Each organ should be placed in a frame of reference by indicating the organs attached to it on either end. For example: "Each vas deferens carries sperms from an epididymis through the inguinal canal, to a point where both vas deferens join with the urethra . . ."

(2) The structures should be described in their appropriate order, beginning at the testis or ovary and progressing to the outside.

(3) The scientific terminology indicated in Figure 15–5 should be employed.

EXERCISE

The following structures are to be identified with the aid of Figures 15–1 through 15–4. Exchange with your neighbors pigs of opposite sexes so that all of you may see both sets of organs. Follow the directions for dissection on page 174.

Make sure you are able to identify on the manikins and chart all the organs on your list. Then label the lines on Figures 15–3 and 15–4.

At the end of this laboratory session, turn in your descriptions of the male and female urogenital organs of the fetal pig.

Male Reproductive Organs

Testes: Produce sperms. Formed in the lower abdominal cavity, they migrate down through the inguinal canal to rest outside of the body proper in the scrotal sac. Paired organs.

Vas deferens: Tubes which carry sperms.

Ejaculatory duct: A small muscular duct formed by the fusion of the vas deferens in the region of the seminal vesicles. Forces sperms from the seminal vesicles into the urethra.

Scrotum (scrotal sac): The sac containing the testes. Attached to the outside of the body ventral to the anus.

Epididymis: A crescent-shaped mass of tubules in which are stored the sperms after they are produced in the testes. Paired organs.

Inguinal canal: A canal connecting the abdominal cavity with the scrotal sac. The spermatic artery, vein, and nerve pass through this region.

Penis: The erectile organ through which passes the lower portion of the urethra. The copulatory organ which allows the sperms to be placed into the vagina of the female.

Seminal vesicles: A pair of small glands at the junction of the vas deferens and the ejaculatory duct. The **prostate gland** is buried between the seminal vesicles.

Penile opening: A tiny opening just posterior to the umbilical cord, through which the penis can be extended. (In pig.)

Bulbourethral (Cowper's) glands: A pair of relatively large glands flanking the urethra just before it turns at right angles to enter the penis. The seminal vesicles, prostate, and bulbourethral glands produce a secretion known as seminal fluid which serves as a liquid vehicle for the sperms.

Spermatic artery and vein: The blood vessels servicing the testes and associated structures.

Spermatic nerve: The nerve which innervates the testes and the surrounding region.

Female Reproductive Organs

The ovaries: A pair of bean-shaped organs located just posterior to the kidneys. Produce the ova or eggs.

Uterus: A Y-shaped organ consisting of the **uterine horns** and a **uterine body**, which is a thickened basal region.

Fallopian tubes (oviducts): Carry the eggs from the ovaries to the horns of the uterus by cilia which set up a downward moving current. Very short in the pig.

Cervix of uterus: A muscular, thickened portion at the base of the body of the uterus. Not visible, but feels hard to the touch.

Ostium (ostia): A pair of funnel-shaped swellings at the ends of the fallopian tubes which receive the eggs as they leave the ovaries.

Vagina: The tube leading from the cervix to the urogenital sinus (vulva).

Urethra: Carries urine from the urinary bladder to enter the vagina at a point about midway between the cervix and the opening to the outside.

Urogenital sinus: The urethra joins the vagina to form a common tube, the urogenital sinus, which opens to the outside. (In the human this tube is reflected back as two pair of folds and is called the **vulva**.)

Genital papilla: A small projection ventral to the opening of the vulva. At its base is the **glans clitoridis**, a region richly endowed with nerve endings and comparable to part of the male penis.

Ovarian artery and vein: Blood vessels supplying blood to the ovaries and surrounding areas.

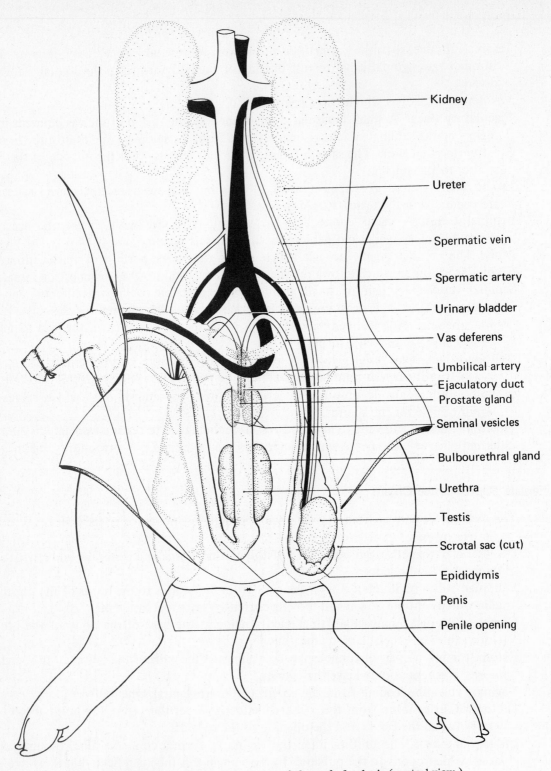

FIGURE 15-1 Urogenital system of the male fetal pig (ventral view.)

Urinary System

Kidneys: Bean-shaped organs which filter nitrogenous wastes from the blood.
Renal arteries and veins: Major blood vessels to and from each kidney.
Ureters: A pair of tubes which carry urine from each kidney to the urinary bladder.

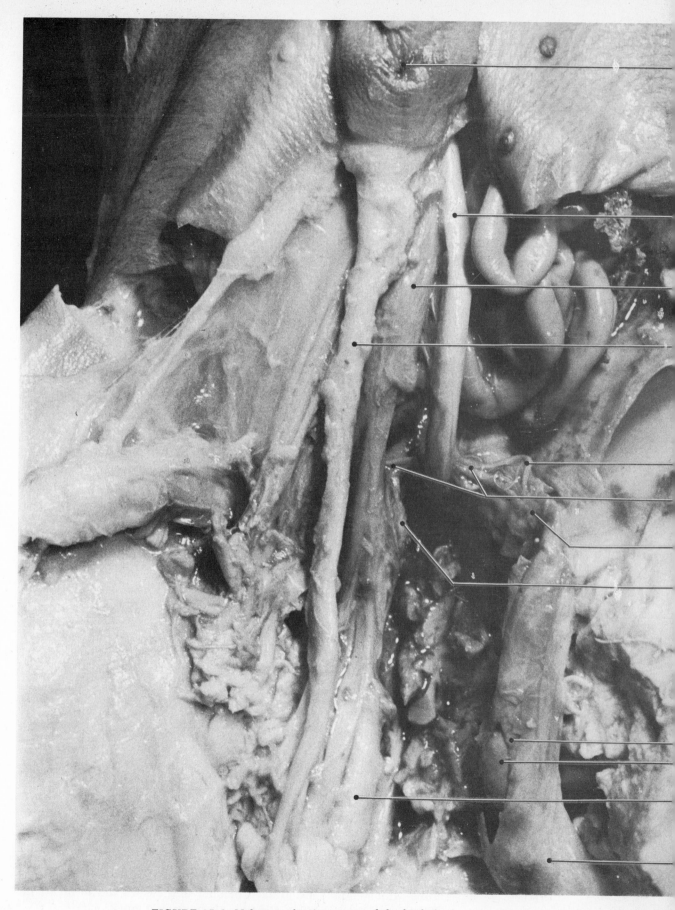

FIGURE 15-2 Male reproductive organs of the fetal pig.

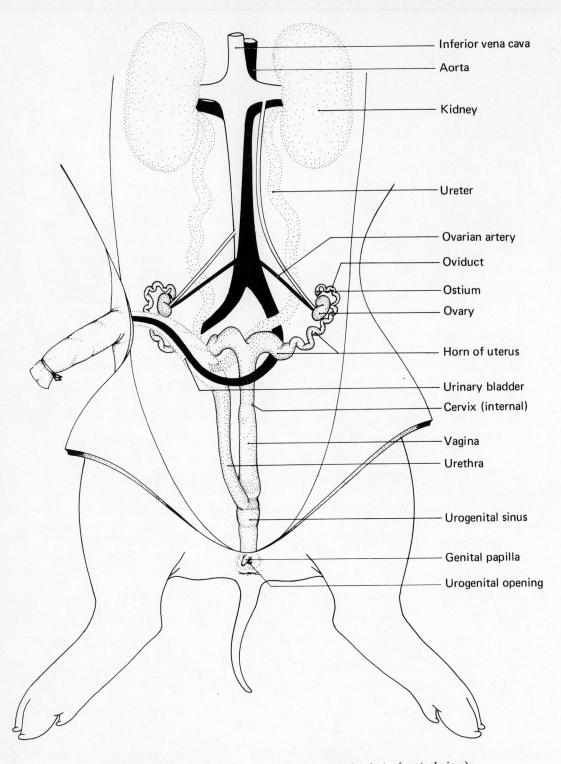

FIGURE 15–3 Urogenital system of the female fetal pig (ventral view).

Urinary bladder: A large deflated sac which seems to form the base of the umbilical
  cord. Stores urine.
Urethra: A single tube which drains the urinary bladder and carries the urine to the
  outside. (Be sure to explain how its course differs in the male and female.)

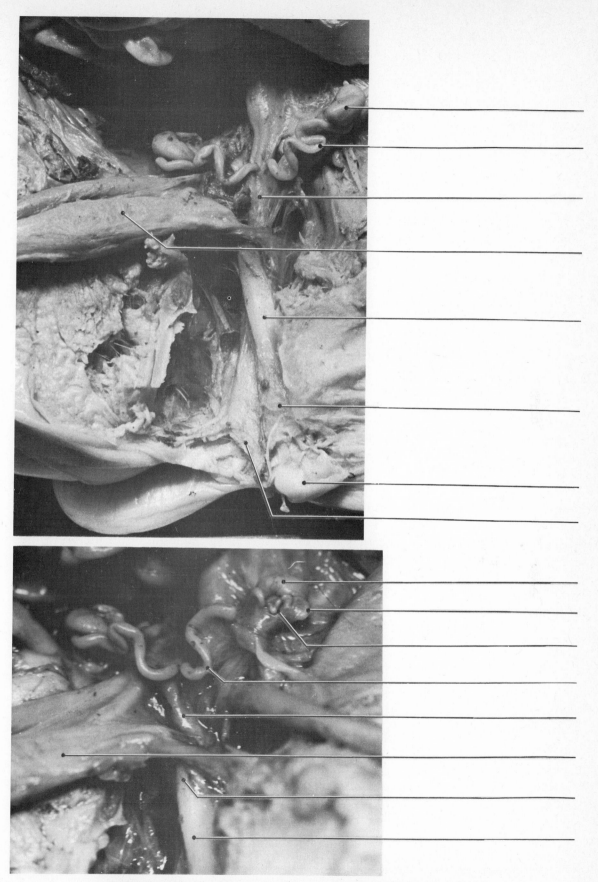

FIGURE 15–4   Female reproductive organs of the fetal pig. The lower
photograph is a closeup view showing ostium and fallopian tubes.

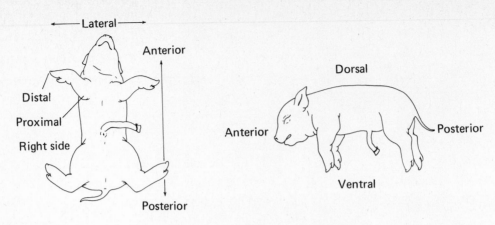

FIGURE 15-5 Scientific terminology.*

Directions for Dissection

Examine the posterior end of your pig (see Figure 15-5). If a sac which appears to be partially inflated is found beneath the tail, you have a male. If instead of a sac a short conical projection is seen just beneath the tail, you are examining a female.

If you have a male, make an incision along the midline of the sac from its origin and below the anus to its most ventral portion. Expose the bean-shaped **testes** within, and cut away the rest of the **scrotal sac**. Clear away the connective tissue surrounding one testis. You are now in a position to trace the remaining male reproductive organs according to the list on page 169.

If you have a female, look for the bean-shaped **ovaries** partly surrounded by their funnel-shaped **ostia**, which are located in the lower abdomen just below the much larger kidneys. Look carefully at each ovary, turning it over gently. You will see a tiny tortuously curved tube extending from each ostium. These are the fallopian tubes or oviducts. In pigs, the fallopian tubes are very small, and they lead to the thin elongated horns of the uterus. (In the human, the fallopian tubes are longer and the horns of the uterus are absent.)

Once you have identified the paired ovaries, their ostia and fallopian tubes, and the horns of the uterus, you are ready to trace the rest of the female reproductive system according to the list of female organs on page 169.

*Note: Proximal and distal are used only when describing parts of appendages; for example, the proximal part of the arm is near the shoulder, while the fingers comprise the distal part. Anterior and posterior are appropriate when referring to parts within the trunk of the body.

VI Homeostasis

16 Inferences from Data

Innervation of Muscles

The scientific method is a formalized reference to the type of thinking which all of us perform many times every day. By analyzing this pattern of thought processes we can see where we can apply scientific thinking and where it may be inappropriate. Some every day examples where "scientific" thinking is appropriate follow.

Most of our behavior is initiated by stimuli received from our environment. It is desirable to perceive these stimuli as accurately as possible, for our response will naturally be dependent upon our version of what we have observed. For example, you may see someone across the street with whom you believe you are acquainted. You may initiate a series of actions such as running across the street and clapping that person on the back, only to find, when he turns around, that your original observation was inaccurate.

When a woman "tries on" many hats, she is experimenting to determine which hat looks best on her. She is testing her hypothesis that one hat will look better on her than all the others.

At the end of the semester, your instructor will be responsible for giving you a grade (that is, his estimate of your performance in his course). He makes his decision on the basis of your scores on several tests, and upon his recollection of your classroom performance. On the basis of these data, your instructor is making an **inference** (a conclusion based on evidence) that your work is "good," "fair," etc.

Today you will be asked to make inferences from data concerning the mechanism or mechanisms which cause and regulate the contraction of cardiac muscle (heartbeat) and striated muscle (voluntary movement).

■ **THE PROBLEM:** *What factors cause and regulate the heartbeat?*

The instructor will pith live frogs, *Rana pipiens*, by inserting a dissecting needle through the foramen magnum and destroying the brain.

1. Can the frog feel pain after the instant when its brain is destroyed? _____

Explain. _____

_____

2. What is the scientist's justification for sacrificing the frog?_____

_____

_____

_____

3. What is the justification for sacrificing the frog in this class? _____

_____

_____

4. Will the frog be dead after it has been pithed? Write a tentative answer to this question below, and submit an essay of approximately 150 words entitled "What Is Life?" or "What

Is Death?" at the beginning of your next laboratory period. _____

_____

_____

_____

THE CONTRACTION OF CARDIAC MUSCLE

Procedure

Obtain a pithed frog at the front table (one frog to each pair of students), and place it, ventral side up, on the wax of your dissecting pan. With your forceps pick up a fold of the loose skin at the midabdomen and cut it with the scissors. Insert the blunt end of the scissors into the incision and cut along the midline up to the underside of the chin; then cut down to the anus between the hind legs. Next, make two cuts at right angles according to Figure 16–1a. Carefully cut through the abdominal muscle and make two flaps as shown in Figure 16–1b. Look for the heart in the middle of the chest cavity. **Remove the heart membrane** (pericardium) if the heart is not clearly visible.

5. Is the heart beating? _____

6. Record the rate of heartbeat for 3 successive minutes.

   1st minute _____

   2nd minute _____

   3rd minute _____

   Average _____

Exercise A

Acetylcholine is a substance found at nerve endings throughout the body. It is also secreted by the **vagus nerve**, a major nerve which originates in the brain. By pouring acetylcholine on the heart we are simulating what occurs when the vagus nerve carries an unusually large number of impulses from the brain. Place 3 drops of acetylcholine solution on the heart.

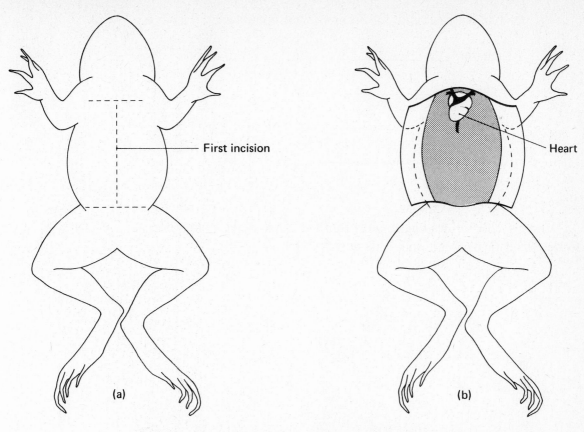

FIGURE 16–1   Ventral view of a frog prepared for dissection.

7. Wait 3 minutes and then record the heart rate for 3 consecutive minutes.

   1st minute _____

   2nd minute _____

   3rd minute _____

   Average _____

Wash the heart by flooding three times with amphibian Ringer's solution, a salt solution isotonic with the frog's cells. This will remove all traces of acetylcholine and bring the heart rate back toward its normal level. *Make sure to wash thoroughly!*

Exercise B

Place 3 drops of adrenaline (epinephrine) on the heart. This substance is secreted by the **adrenal gland** and also by the **sympathetic nervous system.** By pouring adrenaline on the heart, we are simulating what occurs when either the adrenal gland increases its rate of secretion or the sympathetic nerves carry a great number of messages from the brain. If the heart has stopped beating, add 1 drop of epinephrine each minute for about 3 minutes. This should cause it to start beating again. If not, massage the heart with the blunt end of a dissecting needle or pen or pencil.

8. Record the heart rate for 3 consecutive minutes.

1st minute _____

2nd minute _____

3rd minute _____

Average _____

Exercise C

Sever the vessels leading to and from the heart at the points indicated in Figure 8–2 by cutting beneath the heart and at its anterior end.

FIGURE 16-2 Frog heart (ventral view).

9. Remove the heart and place it in a dish of warm Ringer's solution. Carefully observe the heart for a minute and record your observations. _____

_____

_____

10. Record your observations concerning the factors which affect heart rate. Describe each factor and explain *in detail* its effect on the heart rate and activity. _____

_____

_____

_____

_____

_____

_____

CONTRACTION OF SKELETAL MUSCLE

Are the data which you have obtained concerning the innervation of cardiac muscle applicable to skeletal muscle also? To answer this question, turn back to your frog.

Procedure

Cut the skin around the thigh of a hind leg at the point where the leg joins the body, separating the leg skin from the body skin. Grasp the leg skin with a pair of forceps and pull downward, peeling off the skin as one might remove a glove. If this is done correctly, the muscles of the leg should be exposed.

Examine the calf of the leg (below the knee). Insert a dissecting needle between the muscles and slide the needle up and down to remove the covering (fascia) so each muscle is clearly differentiated. Push the muscles aside until you can see the tibiofibula, the fused calf bones.

Examine each muscle of the calf to determine its two points of attachment. Almost every striated muscle is attached at one end to a bone against which it pulls. The bone remains stationary with respect to that specific muscle (although it may be moved by another muscle). The fixed end of the muscle is called the *origin,* while the end attached to the bone which moves is the *insertion* (see Figure 8–3).

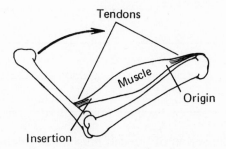

FIGURE 16-3 The attachment of voluntary (striated) muscle.

Note that each muscle is attached to a bone by tendons made of white fibrous connective tissue. Note especially the "tendon of Achilles," which extends from the large rear calf muscle (the gastrocnemius) around the back of the heel and under the sole of the foot.

Trace accurately the origin and insertion of each muscle of the calf of the frog's leg. Make drawings of pairs of these muscles (a pair consists of one muscle on the front of the tibiofibula and one on the back) on the bone outlines provided.

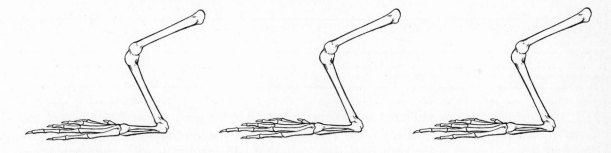

■ **THE PROBLEM**: *By referring to front muscles as A and back muscles as B, hypothesize how the* foot *is moved up and down.*

▢ Hypothesis: _____

_____

_____

Test your hypothesis by freeing the appropriate muscles at their origins and pulling on them. This will simulate their natural action.

Does your hypothesis seem valid?_____If not, revise it.

Your instructor has set up a nerve–muscle preparation at the front table. It consists of the tibiofibula, a pair of calf muscles from a freshly pithed frog, and the nerve which innervates these muscles, the ***sciatic nerve***. Your instructor will stimulate the nerve with electric shocks from a transformer called an inductorium.

11. The instructor will demonstrate what occurs when different sequences of stimuli are

carried through the sciatic nerve to the muscles. Record these data. _____

_____

_____

_____

The instructor will test the hypotheses concerning the action of the calf muscles as volunteered by members of the class. By this direct stimulation of each muscle, the hypotheses can be tested in a more sophisticated manner (that is, in a manner more closely approximating the natural situation) than was possible before.

12. Is your hypothesis valid?_____If not, revise it._____

_____

_____

_____

_____

Are cardiac muscles and striated muscles innervated in the same manner? Members of the class should volunteer suggestions for experiments to test the hypothesis that both types of muscles are controlled in the same way. The instructor will perform these experiments using his nerve–muscle setup.

13. Are both types of muscles similarly controlled? _____

14. In one sentence state the inference you have made from the data, concerning the

innervation of *cardiac* muscle. _____

_____

_____

_____

15. How would you modify this inference when discussing *striated* muscle? _____

_____

_____

_____

16. In the space provided after the following list of sentences place the number 1 for the most comprehensive inference (that which contains a maximum of explicit information) and the numbers 2 and 3 for the second and third most comprehensive, respectively. If several inferences are equally comprehensive, give them the same number. If a sentence is an incorrect inference, write a W (for wrong) after it. If a sentence may be a correct inference but you do not have enough information to be sure, write a question mark next to it.

Remember: *All inferences must be derived only from the data in this series of exercises.*

a. A frog's heart rate will increase if it is frightened.   _____

b. The heart rate of a fish will decrease when it sleeps.   _____

c. A frog's heart rate is controlled exclusively by nerve impulses.   _____

d. A frog's heart rate is controlled by the interaction of various factors.   _____

e. A frog's normal heart rate is more rapid than a human's.   _____

f. A frog's heart rate is controlled by nerve impulses coming from the brain
through the vagus nerve.   _____

g. A frog's heart rate is controlled by chemical factors.   _____

h. A frog's heart rate is not controlled by external factors, but is exclusively
controlled by something inside the heart itself.   _____

i. A frog's heart can beat after removal from its body.   _____

j. A human heart can beat outside of the human body.   _____

k. All animals have hearts.   _____

l.   The rate of the frog's heartbeat is controlled by the brain, chemical factors, and something inside the heart itself.   _____

m.   Cutting the vagus nerve slows the heart rate.   _____

n.   Acetylcholine slows the heart rate.   _____

o.   A frog's heart rate is controlled by something inside the heart and also by nerve impulses.   _____

p.   Striated muscle is innervated exclusively by nerve impulses.   _____

q.   Striated muscle is controlled by all of the same mechanisms as cardiac muscle.   _____

r.   A frog's striated muscle can contract after removal from its body.   _____

s.   Acetylcholine causes striated muscle to contract.   _____

t.   Striated muscle has something inside it which causes contraction, even if all other means of innervation are removed.   _____

17 Construction of an Experiment

Function of Enzymes

An experiment is the test of a hypothesis. It is the means by which a suggested relationship, formulated on the basis of observations, is examined in order to determine its validity. In order to demonstrate that a proposed relationship exists, it is necessary to delineate a cause and its resultant effect.

For example, after many years of research it was finally possible to show that certain viruses cause poliomyelitis. The problem of clearly incriminating these viruses was complicated by factors which masked the cause or the effect. Such factors may be called either "directly relevant" or "not directly relevant" **variables**, depending on whether they actively interfere with the suggested cause–effect relationship or merely confuse the picture by their presence.

A "directly relevant" variable might be an antibody which prevents the disease from developing in spite of the presence of the causative viruses, in which case even though the viruses are introduced into the body of an experimental animal the disease does not appear. Some animals—those not producing enough antibodies and therefore susceptible—become diseased, while others remain healthy. Unless this variable is understood and controlled, the results of the experiment will not be clear.

A "not directly relevant" variable might be other viruses or bacteria found in the blood of the sick animal along with the suspected causative agent. These other organisms may have no relationship at all to the disease, in which case they would be "not directly relevant." However, they may be considered to be "not directly relevant" only after other alternative possibilities have been explored. Perhaps their presence is necessary for the disease to develop, or perhaps they are themselves the causative agent and the experimental viruses are not.

The cause-and-effect relationship can be determined only after the nature of the variables is clearly understood and the pertinent variables are recognized. We will use a simplified plant nutrition experiment for practice in identifying variables.

□ Hypothesis: *Addition of a nitrate fertilizer to the soil will cause plants to grow taller than those planted in nonfertilized soil.*

1. In the space provided on the next page draw an experimental setup and list all of the variables which might affect plant growth. Describe how you would eliminate each variable so that *only the variable relating to nitrates causing increased plant growth* will be the subject of your experiment. DO THIS NOW.

PRELIMINARY INFORMATION

You have been taught that enzymes are important in the process of digestion. They also function in most of the biochemical changes of the body, including the construction of DNA and RNA and in the transfer of energy from certain energy-rich molecules which are not directly utilizable by the body to other molecules which will liberate this energy on demand.

Experimental Setup

<table>
<tr><td align="center">*Variable*</td><td align="center">*Means of Control*</td></tr>
<tr><td>a. _____</td><td>_____</td></tr>
<tr><td>b. _____</td><td>_____</td></tr>
<tr><td>c. _____</td><td>_____</td></tr>
<tr><td>d. _____</td><td>_____</td></tr>
<tr><td>e. _____</td><td>_____</td></tr>
<tr><td>f. _____</td><td>_____</td></tr>
<tr><td>g. _____</td><td>_____</td></tr>
</table>

Thus enzymes are of fundamental importance in many of the chemical reactions which represent the dynamic process we call "life." When digestion occurs, enzymes **catalyze** (accelerate) reactions which result in the breakdown of large food molecules into small "building block" molecules. This breakdown is accomplished by the addition of a water molecule to a large food molecule, causing it to split into smaller components. The enzyme serves as a superstructure into which the pieces of the molecules fit. It does not actually enter into the chemical reaction.

To understand how enzymes work, we can follow an ingested piece of steak to its ultimate destination, the cell.

The steak is broken down into smaller and smaller proteinaceous molecules by enzymes in the stomach and small intestines. These are finally digested to their smallest components, the amino acids. You will recall that amino acids are joined together by **peptide bonds** to form proteins. Therefore, in order to break down the proteins, it is necessary to break the peptide bonds, thus releasing the individual amino acids. Since the peptide bonds represent the loss of a water molecule, it follows that in order to break them a water molecule must be added. In the following reaction it can be seen that digestion is a breakdown of large molecules into smaller ones, accomplished by *adding* a water molecule, whereas synthesis, the build up of smaller molecules into larger ones, is the result of *removing* a water molecule.

Alanine   Glycine   Glycylalanine

2. The equation above shows the synthesis of the compound glycylalanine from the smaller molecules glycine and alanine. Draw the structural arrangement of the reverse of this process, the digestion of glycylalanine into glycine and alanine, showing the exact place on the molecule where hydrolysis (the addition of a water molecule) occurs.

3. If glucose has a formula $C_6H_{12}O_6$, what would be the formula of maltose, a disaccharide made of two glucose molecules formed under the influence of enzymes in plants?

_____ Explain why it is *not* $C_{12}H_{24}O_{12}$. _____

_____

_____

_____

After the steak protein is broken down into amino acids, these small molecules are able to pass through the walls of the small intestines and enter the bloodstream. They then pass into the cells of the body where they are resynthesized into "human proteins" rather than "cattle proteins."

OBSERVATIONS ON THE NATURE OF ENZYMES

Fresh pineapple placed in pure gelatin is soon surrounded by a liquid zone. Perform this demonstration with a small square of gelatin and a piece of fresh pineapple to be found at the front table. While you are waiting for a reaction, go on with this exercise.

4. Gelatin is a protein. Does this liquid zone mean that pineapple contains a proteolytic*

enzyme? _____

5. What other factors might cause the liquefaction of the protein? _____

_____

6. What other source, besides the pineapple cells, might be secreting the proteolytic

agent? _____

_____

■ **THE PROBLEM:** *To explain the liquefaction of gelatin in contact with pineapple.*

□ Hypothesis: *Pineapple releases a proteolytic enzyme.*

The testing of this hypothesis involves two separate elements, so we break it down into subproblems, each of which, when solved, will contribute information toward the final solution.

□ Subproblem 1: *To eliminate other variables which might be producing the effect (the dissolution of the protein).*

□ Subproblem 2: *To demonstrate that the substance produced by the pineapple is indeed an enzyme by showing that it behaves like one.*

The first step in attacking any problem is to outline a plan of action. In formal experiments, this **protocol** or step-by-step program is formulated at the beginning of the study and modified as unforeseen variables appear. Otherwise these might cause the original procedure to become ineffectual through the failure to control newly apparent variables.

Before you can establish a protocol, you must have facts which, seen together, suggest one or several hypotheses. *The following procedures provide the necessary data to enable you to state a valid hypothesis concerning the presence or absence of a proteolytic enzyme in fresh pineapple.*

Proteo- refers to protein; *-lytic* means liquefying and causing dissolution; hence *proteolytic*, breaking down the protein.

Preliminary Data

Procedure I

Place 10 large drops of saliva into a test tube (A) containing 1 ml of starch suspension. Place 10 drops of water in each of two test tubes (B and C) each containing 1 ml of starch suspension. Mix the contents of each tube by flicking the bottom with the forefinger of your left hand while the tube dangles from your right hand.

7. Place 1 drop of iodine solution into tube C. What color does the starch suspension

become when iodine is added to it? _____

When the iodine test is used, this color always indicates the presence of starch.

8. Hold your book up vertically and place tubes A and B against the book. Is it possible

to read what is behind the solution in each tube? _____

9. Set the tubes aside for 20 minutes (go on to procedure II, next page, while you are waiting) and then place them against the books as before. Can you now read through

solution A? _____   Solution B? _____

When large molecules are broken down into smaller ones, the opacity of the solution containing them often disappears and the solution becomes relatively transparent. This can be taken as evidence of enzyme activity and will serve to corroborate your findings using the iodine test below. **Both the transparency and iodine tests have the same purpose.**

10. What can you infer about what has occurred in test tube A? _____

_____

Are you sure?_____ Explain. _____

_____

_____

11. Add 1 drop of iodine solution* to tubes A and B (after waiting the 20 minutes speci-

fied above). Is there any difference in the color of the two solutions? _____ What is

the significance of the difference? _____

_____

*The iodine test is more accurate than the transparency test and should be considered the final arbiter in any apparently contradictory results.

12. Does this add substantiation to the inference you made when you observed the transparency of the solution in tube A? _____ Explain. _____

_____

_____

13. In summary, what do you think has happened to the starch in tube A? _____

_____

_____

Procedure II

Obtain three more test tubes, each with 1 ml of starch suspension. Label them D, E, and F. To tube D add 10 drops of boiled and cooled saliva from the bottle at your table. Into tube E place 10 drops of dilute hydrochloric acid and add 10 drops of unboiled saliva. (**Caution:** *Report to your instructor if any acid spills on you.*)

Place 10 drops of unboiled saliva into tube F and put the tube into the dish of ice at your table. Allow the tubes to stand for 20 minutes and repeat the transparency and iodine tests.

14. Record your data below.

D	E	F
Starch and Boiled Saliva	*Starch, Saliva, and Hydrochloric Acid*	*Starch, Saliva, and Low Temperature*
_____	_____	_____
_____	_____	_____

15. On the basis of the exercise you have just performed, what are some factors which characterize the behavior of enzymes?

a. _____

b. _____

c. _____

d. _____

e. _____

Protocol

You are now ready to construct your protocol or plan for determining whether or not fresh pineapple contains a proteolytic enzyme.

When we stated the problem originally, it was suggested that the solution would be more easily arrived at if we separated the major problem into two less complex subproblems, each of which could be more easily solved than the original problem. These subproblems were

(1) To eliminate variables other than an enzyme as possible causes.

(2) To demonstrate that the substance produced by the pineapple is an enzyme by showing that it behaves like one.

16. List possible causes other than enzymes for the liquefaction of the gelatin. _____

_____

_____

17. Construct an experiment testing each of these possible causes. Describe each experiment and record your results. In your study of a representative enzyme in saliva you used two tests, the transparency and iodine tests. Both had the same purpose, to determine the presence of starch. Using the two tests was a form of insurance to make sure any changes were noticed. Similarly you can use two tests to determine whether or not pineapple contains a proteolytic enzyme. The gelatin block test is often hard to observe. In addition to this test **use small squares of exposed film.** The dull surface is made of a thin coat of gelatin containing a photographic emulsion, which is black. If the gelatin coat is liquefied by a proteolytic enzyme, a clear spot will appear. You will find small squares of exposed film next to the pineapple at your table.

Note: When using the gelatin squares, turn them over so that the pineapple is placed on the underside. The top has been exposed to air and has developed a thick skin.

While you are waiting for a reaction, continue with question 19. Answer question 18 in about 20 minutes, after you have checked the results of these experiments.

DESCRIBE YOUR EXPERIMENTS AND RECORD RESULTS BELOW.

18. As far as you can tell, is the liquefaction caused by a substance other than an

enzyme? _____ Are you sure? _____ Explain. _____

_____

You now have a substantial basis for the hypothesis that enzymes are present in the pine-apple. However, negative evidence (that certain other variables do *not* cause the liquefaction) is not enough to *prove* the presence of enzymes.

19. How can you obtain evidence that a proteolytic enzyme is present in the pineapple? Write a protocol which outlines the appropriate experiments.

Protocol

Experiments

Perform your experiments, *including a control in each one,* and record your data below:

Experiment A. Variable _____ Results _____

Experiment B. Variable _____ Results _____

Experiment C. Variable _____ Results _____

Experiment D. Variable _____ Results _____

Experiment E. Variable _____ Results _____

Conclusion

20. Does fresh pineapple contain a proteolytic enzyme? _____

21. How sure are you of your answer? (*Check one*.) Absolutely positive _____ Fairly

sure _____ Not sure _____ Don't know _____

22. Why? (What is the basis for your decision as to how sure you are of your answer?)

18 Development of a Hypothesis II

Chemical Coordination in Breathing

During previous laboratory exercises, you were called upon to construct hypotheses suggested by data. Some of these hypotheses, because they could not be tested, were actually useless, although they were correctly aimed at bringing as much of the available information as possible into one generalized statement. A hypothesis is a functional statement of a suggested relationship, but it must be testable. Therefore, we must strike a balance between the desire to make broad generalizations and the need to express our generalizations in a form which is amenable to investigation. The hypothesis "all Americans do not want war" errs in two directions. First, it is not testable because it would be impossible to question every person in the United States to determine his views. Second, it seems to be based on wishful thinking rather than on a true evaluation of available facts, since it is doubtful that all Americans agree on anything.

The following hypotheses are all useless. Beneath each is room for you to explain its flaw.

1. Spontaneous generation does not occur for the simple reason that bacteria infect broth by falling into it.

Flaw _____

_____

2. All living things breathe.

Flaw _____

_____

3. The sun has risen every day for all of recorded time; therefore it is sure to appear on all future mornings, forever.

Flaw _____

_____

4. Spontaneous generation, brought about by the combination of methane, ammonia, water, and other substances in the Earth's primeval atmosphere, caused life to appear about 2 billion years ago.

Flaw _____

_____

5. The Rembrandt painting called *The Night Watch* is the most beautiful painting ever made.

Flaw _____

_____

A hypothesis should include the most precise information available. It should be neither too broad nor too narrow, for excessive breadth will either render the hypothesis untestable or will yield such vague information as to make it practically useless. An excessively narrow hypothesis may be so particular as to lose its relevance to any but the most isolated instances.

PRELIMINARY INFORMATION

Cellular respiration and exchange of gases are characteristic activities of all organisms, but only the higher animals have complex mechanisms and coordinating systems which bring about exchange of gases. The term *breathing* in higher animals refers to those body movements which bring air into contact with internal moist membranes. At these moist membranes the exchange of gases occurs by diffusion; oxygen diffuses into the blood and carbon dioxide diffuses out of the blood into the air spaces of the lungs, from which it is expelled out of the body.

Whenever the concentration of any of the constituent gases in the lungs varies, the concentration of that particular gas will vary in the blood in the same manner. For example, if the oxygen concentration in the air goes up, so will the concentration of oxygen in the blood.

By contraction and relaxation of the muscles attached to the ribs, air is caused to move in and out of the lungs. These muscles are activated, at least in part, by nerve impulses coming from the brain or spinal cord. There are nerve endings in tissues. When some nerve endings are stimulated, nerve impulses are initiated which travel to the spinal cord and brain and are relayed to definite organs that respond in some way. *It is known that a change in the concentration of certain substances in the blood may initiate nerve impulses in the brain or spinal cord.*

■ **THE PROBLEM:** *What factors cause an increase in the rate and/or depth of breathing during and after exercise?*

In order to solve the problem a series of observations and experiments are presented. A hypothesis which tentatively provides a solution is to be stated after the first observations. This hypothesis and others will be accepted, discarded, or modified in the light of the facts that are secured from later experiments and observations.

Observations

6. Take a 3 minute record of your partner's breathing rate while at rest.

1st minute _____ 2nd minute _____ 3rd minute _____

7. Your partner should now perform some strenuous exercise for 3 or 4 minutes. Again take a 3 minute record of his breathing rate:

1st minute _____ 2nd minute _____ 3rd minute _____

8. On the basis of these observations and your previous knowledge, state a hypothesis concerning the cause of the increase in the breathing rate.

☐ Hypothesis: _____

_____

Effect of Exercise on Blood Content

An analysis of blood was made before and immediately after exercise, with the results shown in Table 18-1. (Since a comparison of the values before and after exercise supplies the necessary information, the meaning of "vol %" and "mg %" need not be considered here.)

TABLE 18-1 Analysis of Arterial Blood Before and After Exercise

Substances Analyzed for	Before Exercise	After Exercise
Oxygen (vol %)	15	10
Carbon dioxide (vol %)	53	58
Lactic acid (mg %)	5	30

9. Do these data support your hypothesis? _____

10. They should so increase your knowledge that a better hypothesis may now be formulated. In stating this new hypothesis you should recall that the best hypothesis is usually one that interrelates all the facts available, namely, all the facts available in Table 18-1.

☐ Hypothesis: _____

_____

Often a group of facts will give rise to more than one hypothesis. Creative thinkers will formulate not one but several possible explanations for a situation. Thus one measure of your development in the habits of scientific thinking is the number of reasonable hypotheses you are able to invent for a particular group of facts. The following list of statements is presented to broaden your thinking, in order to encompass more possibilities than you may have taken into account.

11. Place a check mark after each of the following statements which (a) is a *logical hypothesis* and (b) could be *useful as a basis for further experimentation* in order to determine the *stimulus for increased rate and/or depth of breathing*. (Make sure both conditions (a) and (b) are fulfilled before checking a statement.) TAKE PLENTY OF TIME TO CONSIDER EACH STATEMENT.

a. Exercise increases the heart rate. _____

b. The concentrations of some substances in the blood change during exercise._____

c. Decrease of oxygen in the blood is the stimulus for increased rate and depth of breathing. _____

    d. Exercise increases the rate and depth of breathing. _____

    e. An excess of carbon dioxide in the blood is the stimulus for increased rate and depth of breathing. _____

    f. The concentration of oxygen in the blood is decreased and that of carbon dioxide increased during exercise. _____

    g. Decrease of carbon dioxide is the stimulus for breathing. _____

    h. A change in the concentration of some of the chemical substances in the blood is the stimulus for increased rate and/or depth of breathing. _____

    i. An increase in lactic acid is the stimulus for increased rate and depth of breathing. _____

    j. Normal continuous breathing depends on the maintenance of a certain level of carbon dioxide in the blood. _____

    k. An increase of oxygen is the stimulus for increased rate and/or depth of breathing. _____

    l. Oxygen was needed by the individual; therefore the rate and depth of breathing increased. _____

    m. An increase in the acidity of the blood is the stimulus for increased rate and/or depth of breathing. _____

12. Which of the preceding statements are illogical hypotheses? _____

13. Why are they illogical? Answer this for each one chosen. _____

_____

_____

_____

_____

_____

14. Which of the statements are not hypotheses, either logical or illogical? _____

Why are these statements not hypotheses? _____

_____

_____

15. On the basis of the information presented in Table 18–1 only, which do you think is

the best hypothesis? _____ Why is it the best? _____

_____

_____

_____

Experiments on the Concentration of Gases in the Blood

Procedure I

The carbon dioxide absorption apparatus will be demonstrated by your instructor. Note that the subject's nose is stoppered so that no air enters his lungs from the outside during the experiment. Instead he will have to rebreathe the air which has entered the apparatus, but the carbon dioxide in it has been absorbed by crystals of calcium hydroxide (which have been changed to calcium carbonate by combining with the carbon dioxide in the air).

16. Count and record the rates of breathing for 3 consecutive minutes.

1st minute _____ 2nd minute _____ 3rd minute _____

17. During this experiment what happens to the volume of the nitrogen in the tank?

(Nitrogen gas is not used by the body.) _____

To the oxygen? _____ To the carbon dioxide? _____

18. What can you infer concerning the quantities of (a) nitrogen, (b) oxygen, and (c) carbon dioxide in the blood at the end of the experiment as compared with those present at the

beginning? _____

_____

19. On the basis of the material presented in this experiment, which of the logical

hypotheses that you checked can be eliminated? _____ Of the remaining list of logical

hypotheses, which ones are still tenable? _____

20. Is your original hypothesis in your list? _____ If not, why have you eliminated it? _____

Procedure II

In order to test the hypotheses still acceptable after procedure I, breathe into a plastic bag as demonstrated by the instructor.

21. Count and record the breathing rates for 3 minutes.

1st minute _____ 2nd minute _____ 3rd minute _____

22. *Describe* any change in depth of breathing if observed. _____

_____

23. During this experiment what happens to the volume of oxygen in the plastic bag?

_____ To the volume of carbon dioxide? _____

To the volume of nitrogen? _____

24. What can you infer concerning the quantities of oxygen and carbon dioxide in the

blood during the experiment? _____

_____

25. On the basis of these two experiments and the data in Table 18–1, choose the best

hypothesis (or hypotheses). _____

26. Why are you now able to eliminate more hypotheses than you could after your

first experiment? _____

_____

_____

_____

Procedure III

Hold your breath as long as possible. Time? _____

27. On the basis of this experiment why do you think you started breathing again? _____

_____

28. Using this fact and all facts previously brought out, state a hypothesis (or hypotheses) which you consider logical in the light of all these facts.

□ Hypothesis: _____

_____

_____

Procedure IV

A subject overventilates by deep and rapid breathing until he feels dizzy. All normal people who overventilate have an involuntary cessation of breathing for about a minute. Analysis of the blood of a person who overventilates shows that carbon dioxide concentration is decreased, oxygen concentration is increased, nitrogen concentration remains the same, and lactic acid concentration increases negligibly.

Re-examine the original list of possible hypotheses.

29. With this additional information in mind, state which hypotheses are in agreement with all the available facts. _____

30. Are any of these hypotheses in any way contradictory to each other? _____

_____

31. What can you *predict* about the control of breathing of (1) frogs and (2) mice from the conclusions you have made in the previous experiments? _____

_____

_____

_____

_____

32. Which of the following judgments would you make about your predictions for

(1) frogs? _____ (2) mice? _____ (3) human beings?_____
    (a) positively correct
    (b) possibly correct
    (c) possibly incorrect
    (d) positively incorrect
    (e) valueless because of lack of sufficient information

33. Describe below (and on the next page) how you could demonstrate that your choice is correct. What aspect of scientific method has led you to your prediction: induction or deduction?

19 Relationship of Data to Solution of a Problem

Some Characteristics of Blood

The purpose of this laboratory exercise is to give you practice in formulating conclusions on the basis of relevant data. It is often possible to draw a tentative conclusion on the basis of inconclusive or partial evidence. However, even though more complete evidence is available, hastily drawn or partial conclusions are too frequently offered.

In this laboratory session, you will be presented with a problem and given the opportunity to accumulate evidence toward its solution. It should be understood that the more evidence you can marshal to support your conclusion, the more substantial it will be. Your task, then, is to see to it that you utilize all the evidence which you can uncover.

PRELIMINARY INFORMATION

Since most parts of the body of a multicelled organism are not in contact with the external environment, a system of vessels (the vascular system) and a circulating fluid (blood) are essential for the maintenance of a proper environment for all living cells in the body. The circulatory system is an expressway for the transport of food to the cells, waste products from the tissues, hormones throughout the body, and respiratory gases. The circulatory system also plays a major role in the regulation of body temperature and in maintaining homeostasis (the internal balance of body processes).

■ **THE PROBLEM:** *Is all blood identical?*

When a human, a frog, or a mouse is injured, red blood pours from the wound. The appearance and behavior of each of these organisms are very different, but their blood appears to be the same. Is it possible that blood is a relatively constant factor among living things despite the fact that most life is characterized by great diversity in almost all physical aspects?

□ Subproblem 1: *What is the physical state of blood?*

Observations

Obtain a sample of your blood by using the following method *(organize material first):*
Clean the ball of a finger by wiping it with alcohol soaked cotton. Puncture the skin with a sterile lancet in the manner demonstrated by your instructor.
Observe the appearance of the blood. Place a drop of it on each of two *clean* glass slides.

On the first slide place a drop of physiological saline solution. The effect of this solution is to dilute the blood. It does not change the appearance or characteristics of the blood cells, since it is isotonic to them. Place a cover slip on the preparation.

On the second slide make a blood smear in the manner demonstrated by your instructor. *Be sure to organize your materials before you begin your preparation;* otherwise you will spend so much time looking for things that the blood will dry before it can be used, necessitating another blood sample.

1. Describe the blood as it appeared on your fingertip. _____

2. Describe the blood in the physiological saline solution as it appeared under low and

high-dry magnifications. _____

_____

3. How does the appearance of the blood in observation 2 differ from that in

observation 1? _____

_____

4. Hypothesize as to why the color no longer appears bright red.

□ Hypothesis:_____

_____

_____

Stain your blood smear with 4 drops of Wright's stain (over a plastic tray or the sink) for 3 minutes. Then flood the slide with 6 drops of Wright's buffer solution for 5 minutes. Wash the slide by placing it in a stream of gently flowing water for a few seconds. Allow it to dry by placing it face-up on a paper towel under a lamp. When completely dry, examine it under low, high-dry, and oil immersion (if available).

5. Can you see any more detail in the red blood cells after staining? _____ Can

you see any other kinds of cells which were not apparent before? _____ How do

they differ from red blood cells? _____

6. Make another description of blood, incorporating all of the information so far

obtained from your observations._____

_____

_____

7. Why were you able to expand upon your first observation? _____

_____

□ **Subproblem 2:** *Is the blood from all organisms the same?*

Observations

Examine a live clam, snail, or insect at the front table.

8. Is it likely that these animals have blood? _____ Why? _____

_____

_____

Examine the exposed, transparent heart of the organism. In order to expose the heart, it was necessary to damage some tissue, causing the animal to bleed.

9. Is its blood similar to human blood? _____ What is the difference? _____

_____

The hemoglobin or red pigment of the blood in many animals enables the blood to carry large quantities of oxygen in a chemically bound state, rather than only in solution. The extra oxygen thus made available permits a high rate of metabolism in the body's cells. Thus, although oxygen molecules fit between the water molecules in liquids (the oxygen molecules fit between the water molecules in a mechanical rather than a chemical relationship), hemoglobin allows the blood to carry sixty times as much oxygen as it could carry if it did not contain red blood cells.

10. What alternative hypotheses can you think of to explain how insects, snails, and clams can survive without this red pigment?

□ Hypothesis I: _____

_____.

□ Hypothesis II:_____.

_____

Consult reference material before your next laboratory period to find out which of the hypotheses suggested above is more correct.

11. Obtain a slide of the blood smear of a reptile, amphibian, or bird. Examine it under low and high-dry magnification. Compare it to the smear you made of your own blood. Are

the two slides identical?_____ Describe any differences you find between human blood and that of these other animals. _____

_____

_____

   12. What is your conclusion to the problem "Is the blood of all animals the same?" (Incorporate all your observations into your answer.)

_____

_____

_____

_____

□ Subproblem 3: *Is the physical state of blood always the same?*

Observations

   Examine the fingertip previously punctured.

   13. In what way is the blood on your skin different from that in your vessels? _____

_____

   14. Is this difference simply due to the fact that the blood has dried, or is there actually an intrinsic change in the physical state of blood when it comes into contact with air? In other words, hypothesize as to whether or not a blood clot is simply dried blood.

□ Hypothesis:_____

_____

_____

   In order to test the hypothesis you have just stated, perform the following experiment with your partner.
   Fill a capillary tubule with your blood in the manner demonstrated by your instructor. Note the exact time you filled the tubule. Your partner should use a watch with a sweep second hand to time the experiment.
   After the tubule is filled, break off a segment approximately a quarter of an inch long at intervals of 30 seconds. Draw the broken piece away from the tubule slowly and look for a change in the appearance of the blood as you break off each piece of tubule. Record the

number of seconds elapsed before a change occurs. _____ seconds.

   15. Is there a difference between a clot and the simple drying of blood? (You may wish

to look at the clot under the low power of your microscope, or under a simple hand lens.) _____

What evidence do you have for your answer? _____

_____

Your instructor will record the clotting time of the members of your class on the chalkboard. Fill in Table 19–1 from these data.

TABLE 19-1 Clotting Time of Students

Initials of Students	Time in Seconds							
	30	60	90	120	150	180	210	270
1.								
2.								
3.								
4.								
5.								
6.								
7.								
8.								
9.								
10.								
11.								
12.								
13.								
14.								
15.								

16. Is there any difference in clotting time? _____

17. Calculate the average clotting time. _____

18. On the basis of the data presented in Table 19–1, is the physical state of blood always constant?_____ Explain. _____

_____

_____

19. Make up a hypothesis concerning the mechanism causing the development of a clot.

□ Hypothesis: _____

_____

Consult reference materials before your next laboratory period to determine the currently accepted theory of clot formation.

□ Subproblem 4: *Is all human blood identical?*

One often hears certain people referred to as "bluebloods," while others are said to have "criminal blood," etc. There is no evidence that human blood varies according to intellectual capacity, wealth, or "personal elegance." However, this does not preclude the possibility that there are differences between types of human blood *which cross personality and racial lines.*

It is known that if the blood cells of one individual are introduced into the bloodstream of another, the donor's cells may be accepted by the body if they are antigenically identical to those of the acceptor. If the cells are different, they will provoke a response similar to that initiated by bacteria and other "foreign proteins" entering the body; they will react with the antibodies found in the blood and be destroyed.

Antibodies are substances carried by the blood. They are manufactured by certain white blood cells and some organs of the body, and they destroy invading organisms or "foreign proteins." This is accomplished in several ways. For example, the cells of bacteria can be "lysed" or dissolved by the antibodies, or their cell membranes can be structurally modified so that they become "sticky" and thus *agglutinate* or clump together, in which state they are more easily destroyed.

At the front table you will find small bottles labeled "anti-A," "anti-B," and "anti-Rh." These contain antibodies *against* red blood cells of types A, B, and Rh. Thus, "anti-A" antibody will cause agglutination of type A red blood cells; "anti-B" will agglutinate type B red blood cells; and "anti-Rh" will agglutinate type Rh red blood cells. If your blood agglutinates with both "anti-A" and "anti-B" antibodies, you have type AB blood. If your blood does not agglutinate with either antibody, you have type O blood.

If your blood agglutinates when in contact with "anti-Rh" antibody, you are Rh positive; otherwise, you are Rh negative. When you record your blood type, indicate A, ʙ, AB, or O type and also your Rh type—for instance, "type A Rh positive."

Observations

Take a clean slide and draw two lines on it with a glass-marking pencil, so that it is divided into three boxes. Write A in the corner of one box (referring to anti-A serum), B in another, and Rh in the third. Place a drop of blood in each box and immediately add a drop of the appropriate antiserum (antibody) to it. Mix with a piece of wooden probe which has not been used before. (Throw the wooden probe away immediately after mixing one drop with it. Why?) On another slide place a drop of physiological saline solution and a drop of blood. Mix. Mark this slide C for "comparison."

Agglutination will be visible to the naked eye as masses of clumped cells separated by clear areas, in contrast to relatively uniform nonagglutinated drops of blood.

20. Record your blood type here. _____

Your instructor will tabulate the number of each blood type in the class on the chalkboard.

21. Calculate the percentages of each type in your class below.

% type O Rh positive: _____   % type O Rh negative: _____

% other types in your class: _____   _____

_____   _____

_____   _____

Below will be found a listing of the original problem and subproblems. Under each sub-problem state your solution and *all evidence which you have accumulated to prove the correctness of your solution.*

■ **THE PROBLEM:** *Is all blood identical?*

□ Subproblem 1: *What is the physical state of blood?*

Solution _____

_____

_____

Evidence _____

_____

_____

□ Subproblem 2: *Is the blood from all organisms the same?*

Solution _____

_____

Evidence _____

_____

□ Subproblem 3: *Is the physical state of the blood always the same?*

Solution _____

_____

Evidence _____

_____

□ Subproblem 4: *Is all human blood identical?*

Solution _____

_____

Evidence _____

_____

You have broken the problem down into manageable subproblems, and you have accumulated evidence which has led you to your conclusions. It is now possible for you to state your solution to the problem in a manner which has the maximum validity. This kind of answer to a problem is the measure of a competent observer in business, industry, or wherever *facts* are important. State your answer to the problem below. Give proof for your conclusions as part of your answer.

■ **THE PROBLEM:**   *Is all blood identical?*

20 Scientific Checks and Balances: Duplication and Quantification

Factors Affecting Circulation of Blood

William Harvey described the circulation of the blood for the first time in 1628. His discovery that the blood flows in a circle from the heart to the rest of the body and back again made possible a rational conception of the economy of the body.

Harvey's explanation of blood circulation had a profound influence on the development of biology. It represented a triumph of the newly emergent experimental method combined with a revolution against the unquestioning acceptance of authority as the source of knowledge. Before Harvey's experimental evidence, the blood vessels were thought to contain a mixture of blood and an undefinable mystical substance called *pneuma*. The pneuma was thought to be a "life-giving" substance. This concept was the intellectual distillation of a series of beliefs held by the ancient Greeks, who thought that life was composed of four elements—earth, air, fire, and water. The man who first invented the concept "pneuma" was the Greek physician Galen, one of the most influential interpreters of Aristotle's biology. Galen's interpretations of Aristotle's biological statements were accepted without question throughout the approximately eight hundred years of the Dark Ages.

Harvey's logical, experimental proof of blood circulation was a profound statement directed against the unquestioning belief in authority and the resultant intellectual stagnation characteristic of the Dark Ages. It was an affirmation of the belief that a rational view of the world, based on accurate observation, might lead to new information which, though contradictory to the beliefs of the Ancients, was nevertheless true. This laboratory exercise is designed to give you an idea of the kind of thinking pursued by Harvey.

PRELIMINARY INFORMATION

An important indicator of the rate at which the blood circulates in the human body is the rate of the heartbeat. From previous experience you can probably recall certain factors which you believe affect the rate of heartbeat. List these factors below.

_____

_____

_____

_____

A scientist, noting that the heartbeat varies in intensity, might conjecture about the causes of this phenomenon. By his natural curiosity, his basic assumption that every effect has a cause, and his acknowledgment that the purpose of science is to search for new knowledge, he might be led to think about factors affecting the heart rate.

His first set of procedures would be aimed at establishing a list of conditions which could result in a modification of the circulation rate. By accumulating these observations (facts), he would lay down a framework from which, he would hope, inductions or generalizations might become evident.

In the same way you are to attempt to formulate a framework for your inductions. It should be understood that your recollections of conditions affecting the rate of heartbeat are inherently subjective and could not be used for further scientific investigation until substantiated. It will therefore be necessary to introduce the set of scientific checks and balances which have been designed to assist the scientist in his quest for objectivity. These are (a) the reproducible nature of the observations and (b) quantitative (relatively exact) descriptions of phenomena.

In other words, if a phenomenon is described as precisely as possible, and if a stated relationship can be demonstrated over and over again, the observer has done all that he can to assure the rest of us that his observations are valid.

1. Does the degree of precision of a description of a set of observations affect your ability

to reproduce them? _____ Explain. _____

_____

_____

■ **THE PROBLEM:** *What are some factors affecting the rate of circulation of blood in various animals?*

□ Hypothesis I: *Exercise accelerates hearbeat in humans.*

2. Make up an experiment testing hypothesis I.*

_____

_____

_____

Control _____

_____

Results (observations) expressed *quantitatively* _____

_____

*The pulse is the swelling of the arteries as they respond to the surge of blood from each contraction of the ventricles. Thus each pulsebeat represents a heartbeat. Your instructor will show you how to take your partner's pulse rate. Why is it bad scientific procedure to take your own pulse?

3. Are your results reproducible? _____ Repeat your experiment in order to demonstrate the reproducible nature of your observations.

Results _____

_____

4. Were both results exactly the same? _____ If they varied slightly (for example,

by 1 or 2 heartbeats per minute), would that invalidate them? _____

Explain. _____

_____

5. Is the hypothesis concerning exercise versus heart rate valid or invalid? _____

☐ Hypothesis II: *Changes of atmospheric temperature do (do not)* (circle one) *have an effect on human heart rate. The effect would be to accelerate heart rate when the the body is heated (cooled)* (circle one).

6. Make up an experiment. _____

_____

_____

_____

Control _____

_____

Observations (quantitative data) _____

_____

_____

Conclusion _____

_____

Your instructor will tabulate the data on the chalkboard.

7. Have you clearly shown the relationship between heart rate and temperature

throughout the whole range of temperatures from freezing to boiling? _____

Is it possible that at very low or high temperatures your results might be reversed? _____

8. How would you modify your experiment (if it were humane to do so) so as to test

your hypothesis fully? _____

_____

_____

A scientist, faced with the problem of being unable to obtain data about the human body because his experiment would subject a human being to harm, would search for a suitable animal to serve as a model. (For our present purposes a model is a substitute which behaves in certain relevant respects like the object to which the data are to be applied.)

There is an inherent danger in this approach, since the animal used as a model may not respond to the experimental stimulus in the same manner as a human.

It is possible to divide the animal kingdom into two major divisions, using the blood temperature as a frame of reference. Mammals and birds are homeothermic (warm-blooded), while reptiles, amphibia, fishes, and all invertebrates are poikilothermic (cold-blooded). Homeothermic animals maintain a constant blood (and body) temperature except when exposed to extreme cold or heat. The temperature of poikilotherms is similar to that of the environment.

Suppose that the scientist chose the *Daphnia* as his experimental animal. *Daphnia* is a minute crustacean often found in freshwater ponds. It is especially suited for experiments dealing with heart rate because it has a transparent exoskeleton and sometimes has red blood (unusual in crustaceans, which generally have transparent or bluish blood).

We will attempt to determine if the data obtained by using *Daphnia* are applicable to the human body.

□ Hypothesis III: *When a* Daphnia *is placed in a cold environment its heart rate will increase (decrease)* (circle one).

Experiment

Put a small drop of petrolatum on the bottom of a syracuse dish. Using an inverted medicine dropper pipet, pick up a *Daphnia* from the culture at your table and drop it on the petrolatum. Tilt the dish so that the water runs off and the *Daphnia* is left embedded on its side in the petrolatum. Use an artist's brush for adjusting the *Daphnia* if necessary. Pour pond water into the dish until it covers the *Daphnia*. Prepare your experiment and make your observations using a binocular dissecting microscope. Record the rate of heartbeat of your *Daphnia* at room temperature for 3 consecutive minutes. Divide the total number of beats by 3 to obtain the average. The position of the heart can be seen in Figure 20–1.

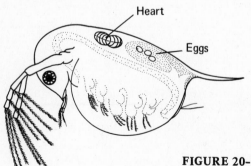

FIGURE 20–1   Position of the heart of the *Daphnia.*

Place your dish on the ice in the metal tray at your table. When the temperature of the water in the dish descends to 17°C (62°F), obtain a 1 minute recording of the heart rate. Do this again each time the temperature is reduced by 6°C.

9. Record your observations. (*Note:* You may not be able to get the water down to 0°C. Record data for the lowest temperature achieved.)

Heart rate at room temperature   23°C _____

Heart rate at temperature   17°C _____

Heart rate at temperature   11°C _____

Heart rate at temperature   5°C _____

Heart rate at temperature   0°C _____

Plot a graph on the paper on page 131 to demonstrate clearly the relationship between temperature (horizontal axis) and number of beats per minute (vertical axis). Use 60 spaces for each axis. Each space on the vertical axis will represent 3 beats (estimate above 180 beats/min). Each space on the horizontal axis will represent ½°C. The left-hand lower corner of the graph will begin at 0 beats going vertically, and 0°C going horizontally.

10. On the basis of the data, what is your conclusion concerning the effect of

temperature on the rate of heartbeat in *Daphnia*? _____

_____

_____

How could you use a warm water bath (consisting of a finger bowl of warm water from the tap into which the dish with the *Daphnia* is placed) to corroborate your findings? Set this up; while waiting for results, go on to question 11.

11. Should the scientist be satisfied with the above data and propose the theory that the

heart rate of *all* animals is always depressed by a decrease in temperature? _____ Why

or why not? _____

_____

_____

_____

To determine whether or not the effect of temperature on the heart rate of animals is accurately represented by the *Daphnia,* the scientist repeats his experiments using other animals. His records of the relationship between mouse, dog, turtle, and frog heart rates and temperature are reproduced in Table 20–1.

TABLE 20-1   Rate of Heartbeat of the Frog, Dog, Turtle, and Mouse as a
Function of Temperature

(Hearts are isolated)

	$29°C$	$23°C$	$17°C$	$11°C$	$5°C$	$0°C$
			Beats per Minute			
Frog	55	45	30	21	6	0
Dog	59 (at 27° rate is 20; at 26°C rate approaches 0)					
Turtle	34	19	9	5	2	0
Mouse	208	129	71	30	0	0

Plot the data in Table 20–1 on the same graph with *Daphnia* heart rate, using a dotted line to represent the mouse data, a red line for the dog, a blue line for the turtle, and a pencil line (or pen, if a pencil was already used) for the frog.

12. Put a T after those alternatives which are valid interpretations of the data on your graph. Use F for false interpretations, and a question mark for statements for which there are insufficient data.

a. The data show that both mouse and *Daphnia* heart rate varies inversely as
the temperature. _____

b. When subjected to decreasing temperatures, mouse heart rate goes up while
Daphnia rate goes down. _____

c. Mouse heart rate is not affected by changes in temperature. _____

d. As the temperature increases from room temperature, *Daphnia* heart rate
increases. _____

e. At low temperatures, both mouse and *Daphnia* heart rates slow down, but
do not stop. _____

f. Mouse and *Daphnia* heart rates are the same at 11°C. _____

g. Human heart rate stops at low (5°C) temperatures. _____

h. Frogs, dogs, and turtles respond similarly to a decrease in temperature. _____

i. All poikilothermic animals have a heart rate response which is directly
related to the temperature of the environment. _____

j. The heart rate of the poikilothermic animals in our experiment is inversely
related to the temperature of the environment. _____

k. Some homeothermic animals die before they are cooled down to 2 or 3°C,
while some poikilotherms are still alive at that temperature. _____

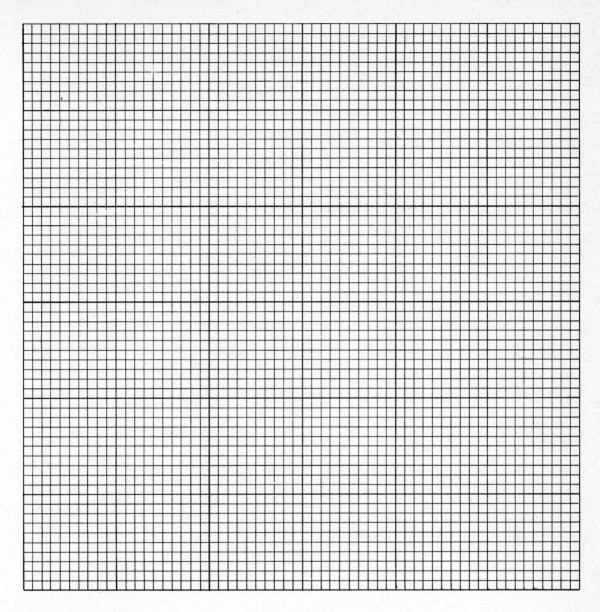

CONCLUSION

13. Make a statement incorporating as much information as you can about the relation-

ship between heart rate and temperature in animals. _____

_____

_____

_____

Your statement is a simplified version of a theory. It represents a generalization formulated on the basis of several validated hypotheses.

218   Scientific Checks and Balances: Duplication and Quantification   [Ch. 20

REVIEW QUESTION

1. Explain the meaning of the terms *duplication* and *quantification*. Next to your explanation describe the role of each in science.

    a. Duplication: _____

_____

_____

    b. Quantification: _____

_____

_____

21 The Ability to Collect and Organize Facts

Behavioral and Morphological Aspects of Reflexes

Much of the behavior of lower animals consists of inherited activities. For example, a spider kept isolated from others of its kind will still spin a web characteristic of its species. A baltimore oriole kept in a robin colony will still make an oriole nest when the stimulus for nest building is presented.

Humans, too, perform certain activities in common with one another which are capable of being evoked regardless of cultural influences. This laboratory exercise is devoted to those activities, called **reflexes**, which are shared by all humans.*

PRELIMINARY INFORMATION

The following exercises are designed to supply facts for you to relate to one another to form a pattern. It is in the nature of science to assume that a pattern exists among closely related phenomena. This pattern must be discovered and stated in the form of a generalization so that the physical causes governing that class of interrelationships can be understood.

When the pattern relating a series of phenomena begins to be apparent, it is stated as an hypothesis or tentative explanation. This hypothesis is an endeavor to peer through clouds of irrelevant information to discern the outline of a truly related system of phenomena. When a detective views the scene of a crime, his senses are literally bombarded with impressions (facts). He sees an open window, an overturned chair, a clock, an unfinished meal, a sleeping cat, a spot of red on the carpet. What do they all mean? Which are relevant and which are not? What picture will be revealed if the facts which are related are properly put together?

■ **THE PROBLEM:** *What is the relationship between the anatomy of the nervous system and the reflexes?*

PART I. OBSERVATIONS OF HUMAN REFLEXES

Students will work in groups of three during this laboratory period, as designated by the instructor. Laboratory reports will be prepared and submitted *independently*. Student 1 will be the observer and recorder, student 2 the experimenter, and student 3 will be the subject.

*It should be understood that reflexes, in general, are much less complex than web spinning or nest building, which are governed by "instincts," sometimes defined as sequentially arranged reflexes.

Exercise A

1. Shine a bright light into the eye of the subject.* Observe the response of the pupil of the eye as the light is alternately applied and withdrawn.

Observation _____

_____

2. The subject should now concentrate on preventing the response. Again apply the light.

Observation _____

3. Can this response be controlled by the subject? _____

Exercise B

The experimenter will stand behind and to the left of the subject, and the recorder will sit directly across the table from the subject. While everyone is getting into position, the experimenter will dip one end of a 12 inch piece of rubber tubing into the alcohol-and-cotton dish at the center of the table. He is to wave the hose in the air for a few seconds to let it dry, and then place the sterilized end in his mouth. The other end should be held about three-quarters of an inch from the left eye of the subject. The experimenter should then position himself to the left and slightly behind the subject.

4. At a signal from the observer, blow a short (and gentle) burst of air into the subject's eye. Repeat this several times to be sure the first observation was no accident.

Observation _____

_____

5. The subject should concentrate on preventing the response. Apply the stimulus again.

Observation _____

6. Can this response be controlled by the subject? _____

Exercise C

The subject will cross his left leg over the right, allowing it to hang limply in a relaxed state. The experimenter will find the depression just below the patella.

7. Hit the tendon found there with the side of the hand or a rubber mallet.

Observation _____

*The response will be enhanced if he closes his eyes for about 30 seconds before the light is applied.

8. The subject should concentrate on preventing the response. Apply the stimulus again.

Observation _____

9. Can this response be controlled by the subject ? _____

Exercise D

Place the thumb of the left hand firmly on the tendon of the biceps muscle of the subject by pressing hard just above the depression of the elbow joint. The tendon will be easily found if the subject "makes a muscle" and then lowers his arm to an outstretched position.

10. Now, as the arm is outstretched and relaxed, the experimenter will hit his own thumb with a rubber mallet. The effect of the blow will be transferred to the tendon of the subject's biceps.

Observation _____

_____

11. The subject should concentrate on preventing the response. Apply the stimulus again.

Observation _____

12. Can this response be controlled by the subject? _____

Exercise E

13. Place a drop of 10 percent acetic acid on your tongue (acetic acid is the active ingredient in vinegar; it is harmless in weak concentrations). After recording the data, rinse your mouth.

Observation _____

14. The subject should concentrate on preventing the response. Apply the stimulus again.

Observation _____

15. Can this response be controlled by the subject? _____

Interim Conclusions

16. What do all of the above reflexes have in common? _____

_____

_____

17. Can all reflexes be controlled by the higher nervous centers (that is, can all reflexes

be consciously controlled)? _____ What is the significance of this fact in terms

of the source or origin of the reflex response? _____

_____

18. The instructor will tabulate the results of the exercises on the chalkboard. Fill in the chart below with the number of subjects who responded positively and negatively to the stimuli.

Reflex	Number of Positive Responses	Number of Negative Responses
Pupil		
Eye wink		
Patella		
Biceps jerk		
Salivation		

19. Are all reflexes elicited with the same ease? _____

20. Explain why certain reflexes were uniformly exhibited and others were not. (Your limited knowledge at this stage of the study will of course make your explanation a tentative

one.) _____

_____

_____

_____

There are many reflexes exhibited by humans. Some representative examples are:

Accommodation–pupil reflex: The pupil of the eye contracts when the shape of the lens is changed to adjust for near vision.

Plantar reflex: When the sole of the foot is stroked from the heel to the base of the small toe, the toes curl.

Sneeze or cough reflex: Irritation of the mucosa of the respiratory passages causes deep inspiration, followed by a closure of the vocal cords which do not open until the expiratory movement is under way. This causes an explosive movement of air which sweeps the offending object out of the respiratory passages.

Aortic reflex: The inner membrane of the aorta is capable of responding to increases in blood pressure, resulting in a decrease in the heart rate and, as a consequence of a slower heartbeat, a lower blood pressure.

21. The reflexes you have studied so far are called *simple reflexes*. Make a generalization about the characteristics of simple reflexes. _____

_____

_____

PART II. THE ROLE OF THE NERVES

Examine the drawing of a cross section of the spinal cord (Figure 21-1). Note that the spinal cord consists of neurons and parts of neurons which carry impulses to and from the brain and regions of the spinal cord. The parts of the nervous system associated with the reflex are:

The receptor: This is an organ which contains neurons specially differentiated to convert changes in the environment (stimuli) into nerve impulses. Light, sound, chemicals, pressure, heat, cold, etc., are aspects of the environment capable of initiating nerve impulses in their special receptors.

The sensory or afferent neuron: This nerve cell carries the impulse from the receptor toward the spinal cord.

The association neuron: This neuron transfers the nerve impulse from neuron to neuron. *Some association neurons carry impulses to the brain,* which consists almost exclusively of association neurons.

The motor or efferent neuron: This neuron carries the nerve impulse from the spinal cord toward the *effector.*

The effector: This is either a muscle or a gland which converts the nerve impulse into a physical or chemical *response* to the original stimulus.

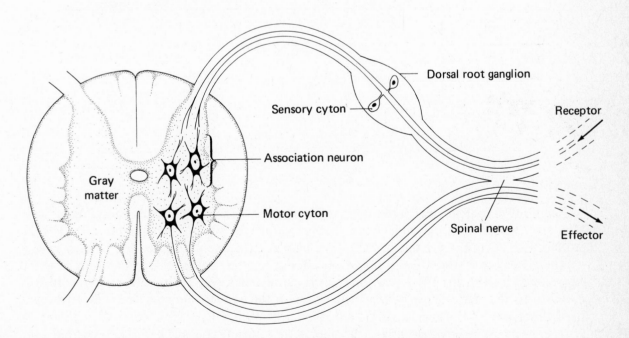

FIGURE 21-1   Cross section of mammalian spinal cord.

Obtain a slide of a cross section of spinal cord from the front table. Examine it under low and high-dry powers. You should not have difficulty finding the cytons of the association and motor neurons in the butterfly-shaped gray matter portion of the spinal cord. Note that the sensory cytons are found *outside* the spinal cord in the *dorsal root ganglia* of the spinal nerves. Examine the slide.

28. Note that the upper portion of the "butterfly" is much narrower than the lower.

Does this have anything to do with the sensory cytons? _____ Explain. _____

_____

_____

29. Using Figure 21-1 and the knowledge you have obtained from your observations of human and frog reflexes, list the neurons involved in the performance of a simple reflex.

    a. Receptor **b.** _____ **c.** _____ **d.** _____ **e.** Effector

30. A classical example of a simple reflex is the withdrawal of one's hand upon contact with a painful stimulus (a hot stove, for instance) *before* any sensation of pain. Explain, by describing neural pathways: **a.** How is it possible for the hand to withdraw automatically *before* the sensation of pain is felt? **b.** What neurons are involved so that soon after the response we *do* feel pain?

    a. _____

_____

_____

    b. _____

_____

_____

Exercise G

Reassemble your original three member team as in the eye-wink reflex study. Again seat the observer across from the subject, with the experimenter slightly behind and to the left of the subject, and with the opening of the rubber hose about ¾ inch from the subject's left eye.

This time your instructor will ring a bell, after which the experimenter will say to himself "one second," and then blow a burst of air into the subject's eye. The observer will record the eye winks on the chart below by placing a check in the appropriate space whenever the eye wink is observed.

There will be fifty consecutive trials, so make yourself comfortable. The experimenter should remain in a more or less fixed position so as not to distract the subject by extraneous movements. (Do not raise and lower the hose with each burst of air.)

The subject should read a book so as to take his mind off the purpose of the experiment. Hold the book in such a way that the observer can see your eyes.

Your instructor will begin with three practice trials.

Trial	Bell	Air		Trial	Bell	Air
1.				26.		
2.				27.		
3.				28.		
4.				29.		
5.				30.		
6.				31.		
7.				32.		
8.				33.		
9.				34.		
10.				35.		
11.				36.		
12.				37.		
13.				38.		
14.				39.		
15.				40.		
16.				41.		
17.				42.		
18.				43.		
19.				44.		
20.				45.		
21.				46.		
22.				47.		
23.				48.		
24.				49.		
25.				50.		

Interpretation of the Data

31. Record the number of the trial which starts a *consistent* sequence of eye winks at the

bell. _____ Your instructor will record on the chalkboard the data for all groups. What is the lowest number of trials recorded before consistent responses to the bell were evident?

_____ What generalization can you derive about the number of trials necessary to cause

the behavior? _____

32. Did all groups report a consistent series of winks at the *burst of air?* _____

Explain why or why not. _____

_____

33. Did all groups report a consistent series of winks at the bell? _____ Give

a tentative explanation as to why or why not. _____

_____

_____

_____

34. What can you say about the number of trials needed to establish the eye wink after

the burst of air as compared to the number required after the bell? _____

_____

35. Does this fit in with your explanation as to why different numbers of trials were

necessary to evoke the winking behavior after the air versus after the bell? _____

_____

_____

36. In what way does the exercise just finished differ from the original eye-wink study

(Exercise B)? _____

_____

_____

37. Is a simple reflex involved? _____

38. What part of the central nervous system (the brain and spinal cord) is *directly* involved in a simple reflex? _____

39. Which studies have you performed which prove this? _____

_____

40. What part of the central nervous system, not usually involved in the simple reflex, is involved in the response when the eye winks at the sound of the *bell*? _____

41. The type of behavior exhibited when the eye winks at the sound of the bell is called a *conditioned* reflex. List each type of neuron and organ of the nervous system involved in a conditioned reflex.

42. Explain the differences between a simple and a conditioned reflex in terms of the neurons and organs involved. _____

_____

_____

_____

43. Explain the association between the stimulus which evokes the simple reflex response and the stimulus which evokes the conditioned response. _____

_____

_____

_____

44. Is there any difference in the conditioned reflex response versus the simple reflex

response? _____

45. In all examples of *conditioned* reflexes must there first be a *simple* reflex? _____

Explain. _____

_____

_____

_____

REPORT SHEET   Name _____

Instructor's name _____

1. Review this entire laboratory exercise. Make a list for yourself of the data obtained, keeping in mind that you are attempting to extract broad meaning from a composite of the separate pieces of information. Your generalization (or induction) will be a kind of jigsaw puzzle of observations. Its value will depend on how well you fit the pieces together.

2. State your conclusion to the problem. It should include all the pertinent data in your list. (*Reminder:* The problem is "What is the relationship between the anatomy of the nervous system and the reflexes?")

22 Perception and Empiricism

Nervous System: Sense Organs

That science is based on observation has, it is hoped, become apparent. This statement has a direct impact upon the lives of all individuals, not only scientists. Is it not true that all our activities from the moment the alarm clock wakes us to the moment we lose consciousness in sleep are responses to stimuli which are perceived by our senses? Is it not true in large part that our impressions of people are in reality composite responses to stimuli evoked by the appearance (visual stimuli) and the nature of the conversation (sound stimuli) characteristic of the individual? Our sense organs are our "window on the world," our only contact with our environment. They are the means by which all higher animals are able to locate their food and escape their enemies.

1. Are plants able to respond to stimuli?_____ Explain and give examples.

_____

_____

2. Are lower animals such as insects, jellyfish, and protozoans able to respond to stimuli?

_____ Explain and give examples. _____

_____

The success which science has achieved in influencing our lives rests, at least in part, on the ability of the scientist to base his thinking on **fact**. By definition, a fact is a statement about something which is independent of the mind and sense organs of any one individual. It is a concept of reality which transcends the weaknesses inherent in the observations of any single perceiver.

3. How does an observation become a fact? _____

_____

We have examined the difference between fact and nonfact for the scientist. We have shown the value of the use of fact in scientific discovery. What is the role of fact in our daily activities? Can we benefit from a clearer picture of reality when we formulate our daily decisions? For example, did you have a factual picture of college life before you decided to go to college? Did you know the advantages of this college over others in this vicinity? Would it not have been desirable for you to have visited several colleges to observe campus life before you made your decision?

4. What activities should a prospective college student engage in, in order to determine whether or not college is right for him, and, if it is, which college is best for his purpose?

_____

_____

_____

_____

P A R T I. P E R C E P T I O N O F S T I M U L I

Perception, for the individual, may be broken down into a series of phenomena wherein an aspect of the environment, the **stimulus**, is incorporated into the framework of experiences of the perceiver. A discussion of the nature of perception, beginning with the stimulus, follows.

The Stimulus

The causation of all perception is the stimulus, which can be defined as any change in the environment capable of causing excitation and activity of the sense organs. A deer moving through the forest produces stimuli by disturbing the air when it steps on a twig. This displacement of air molecules is picked up by our ears as sound. We see the deer when it moves because movement is a *change* in our environment of motionless trees. If the deer were to lie still among the leaves covering the floor of the forest, most of us would not be capable of responding to its presence because the change would not be great enough to cause a response in our untrained senses. An experienced woodsman would probably detect the difference (change) in the pattern of light rays reflecting from the deer and from the surrounding foliage, and thus would be aware of its presence.

Each of our organs has a threshold below and above which a change in the environment is undetectable. A dog will respond to the blowing of a whistle pitched at 20,000 vibrations per second, while the same sound will not constitute a stimulus for us. Bats can locate objects by bouncing supersonic whistles off them. An eagle can see a fieldmouse moving in the grass from a quarter of a mile away. The mouse is sometimes capable of escaping the eagle by hearing the sound of its wings beating against the air, or by seeing its shadow.

Exercise A

Note: Perform Exercise B before tabulating the data for Exercise A.

Your instructor will use a signal generator to produce sounds of varying pitches. Record the number of each signal to which your ears respond on Table 22-1. The instructor will ask you to compare the range of your hearing with that of your classmates.

5. Is everyone capable of responding to sound stimuli at the same level? _____

Exercise B

The instructor will pass out two sets of paper strips coated with various substances

TABLE 22-1 Audibility of Sounds Which Differ in Pitch

Number of Signal	Check If Audible	Pitch, cycles per second (to be announced by instructor after all signals have been produced)
1.		
2.		
3.		
4.		
5.		
6.		
7.		
8.		
9.		
10.		
11.		
12.		
13.		
14.		
15.		

including phenylthiocarbamide (PTC). Record any taste response you may have, for example,

bitter, sweet, sour, salty. Strip 1 _____ Strip 2 _____ The instructor will tabulate the responses of the class on the chalkboard.

6. Is the response to the chemicals on the paper similar in all persons?_____ Your instructor will briefly discuss the cause of this variability, or you can find out for yourself by consulting a textbook on genetics.

Exercise C

Your instructor will show some colored slides. Record the number of each box under the letter of the appropriate color. Your instructor will accumulate the data after you have filled in the chart.

7. Do all persons see the same colors? _____

TABLE 22-2 Color Vision

	Color A	Color B	Color C
Number of Box			

Exercise D

Take two pins and a ruler and perform the following investigation. Have your partner close his eyes. Touch the tips of the two pins to the back of his hand. Begin with the pins about ¼ inch apart. Gradually move the pins further apart. Measure the distance between the pins when your partner indicates that he feels two sensations instead of one. Repeat this test on his lips, fingertips, forehead, and the back of his neck. By using only one pin instead of two, test whether your partner is reporting valid information *or is imagining what he feels.*

8. If you find that your partner is imagining that he feels two pins when you apply only

one, should you immediately assume that he is deliberately misinforming you? _____

Explain. _____

_____

This test demonstrates a very significant aspect of perception: the involvement of past experience in the formulation of the percept or image.

9. Record the degree of sensitivity of the portions of the body tested (1 is most sensitive, 5 is least sensitive).

Back of hand _____

Lip   _____

Fingertip   _____

Back of neck _____

Forehead   _____

Consult reference materials and be prepared to explain why this variation in sensitivity occurs.

Conclusion

10. Do some sense organs vary from person to person in their ability to perceive stimuli?

Use the observations accumulated in the four preceding exercises to support your induction.

_____

_____

_____

The Percept

The percept is the image (sound, touch, etc.) formulated in the mind by the integration of the nerve impulses arising from the stimuli impinging on the sense organ and by the unconscious tendency to relate such stimuli to past experience.

When the stimuli are interwoven (unconsciously) with past experience the resultant whole concept is called a *cognition*. The cognition is, in effect, the placing of isolated stimuli into a framework of past experience so that they become part of the total knowledge possessed by the observer. A classical example of cognition is the placing together of a group of sticks into what is known to us as a "chair." We perceive the group of sticks as a unit (chair), not merely as a complex of separate entities. A Martian, not having our previous experience, might perceive the group of sticks as such, and not as a functional whole (something to sit on).

The percept is formulated as the result of the interplay of the factors listed in Table 22-3. Next to each will be found the possible adverse influence which may result from their introduction into the perceptual process.

TABLE 22-3.   **Factors Causing or Influencing Perception and Their Effects**

Perceptual Factors	Effect on Formation of the Percept
1. The stimulus	If the stimulus is almost beyond the capacity of the sense organs to respond to it, only part of the stimulus may be responded to, causing a percept which does not reflect the true stimulus. A radio of poor quality may not reproduce the high notes of the piccolos and the lows of the bass drum. You might conclude, after listening to a performance on this radio, that the music was poorly played. Your sense organs can distort the nature of a stimulus in a similar manner.
2. The sense organs	The sense organs of each individual differ as to their capacity to respond to the stimulus. If there is a defect in the sense organ it will distort the nature of the stimulus. A person with astigmatism might see this ⌡ instead of this ↑ Some people are born with sense organs which are more sensitive than others. For example, certain individuals have what is known as "true pitch"; they can differentiate by ear the exact pitch of a note, whereas others can never cultivate this skill. Visual acuity is another variable property (of the sense organs). Later on during the period, determine the keenness of your vision with a Snellen Vision Test hanging at the front of the room.

TABLE 22-3 (*cont.*)

Perceptual Factors	Effect on Formation of the Percept
3. The mind, memory, integration, etc.	There is an innate tendency for the mind to relate the stimulus to other experiences which are recorded as memories. If the particular stimulus is not part of the fabric of the observer's past experience, there is still a tendency to place the observation into some context. In such cases it is likely that the percept—the final picture—will be distorted so as to make the original stimulus fit into the experience framework of the perceiver, so strong is this tendency to interrelate stimuli.
4. The phenomenological report	In view of the information previously presented, it is highly improbable that the final image or percept in the mind of the observer is an exact reproduction of the original stimulus. Too many interferences lie between the original stimulus and the final percept. The nature of the stimulus, the nature of the sense organs, and the nature of the integrative apparatus of the mind make it unlikely that what appears in your mind when you look at something will be an exact reproduction of what actually exists. We must therefore understand that no report of a perception can be *perfectly* accurate. The term *phenomenological report* has been devised to refer to reports of perceptions incorporating the above mentioned influences. The scientist, however, must deal with facts, that is, true representations of reality. How does the scientist insure that what he is observing is not colored by the perceptual process? One way is to insist on a quantitative statement of what is perceived. The scientist will report a sound as 440 cycles per second (pitch) at 30 decibels (loudness). He has described his perception in terms which are as accurate as possible. You or I would refer to a sound as A-natural if we wanted to be accurate. This is our way of describing a sound perception in quantitative terms.

If we wish to communicate a perception, we can (a) describe our emotional response to it, (b) describe it in general terms, or (c) describe it as specifically as possible. For example: (a) It was a beautiful piece of music. (b) It was a choral symphony. (c) It was a choral symphony in the key of B-flat, in four movements, scored for tenor, soprano, alto, and bass, violins, tympani, etc.

The scientist would find most useful the specifics of the last description, which fulfills the major requirements of perception in science—detailed description and accuracy. For it must be borne in mind that even if a phenomenon is described in such detailed and accurate terms as to permit others to perceive it, the *requirement of empirical verifiability will not have been met until others have actually perceived it.*

In our everyday lives, as in the concert described above, we often confuse perception (a) or perception (b) with perception (c). It is always desirable to delimit in one's mind the boundaries of the various kinds of perception.

A description of a stimulus (the report of a perception) is called a **phenomenological report** if it includes any influence or modifications due to the observer's perceptual apparatus. *All perceptions we make are to some degree phenomenological,* since the very nature of our sense organs makes it almost impossible for us to avoid distorting to a greater or lesser extent the images we perceive. We must make every effort to exclude the weaknesses of the perceptual process by being constantly aware of the pitfalls discussed above. This awareness will enable us to approach a description of the stimulus *as it really* is. This ultimate, and *impossible*, perfectly accurate description is called a **physicalistic report.** While its achievement is unlikely, the closer we approach this type of relationship with our environment, the closer we come to facing reality as it really is; and the more able we are to cope with it.

Exercise E

Take a piece of wire mesh and hold it about six inches from your face.

11. State briefly what you see. _____

_____

12. Check with several neighbors. Have they all seen exactly what you have seen? _____
Now focus your eyes on the mesh itself. Can you see objects at the end of the room (through

the mesh) clearly? _____ Now focus on an object across the room, still keeping the

mesh in front of your eyes. Can you see the object clearly? _____ Can you see the

mesh clearly? _____

13. Describe what occurred when you focused first on the mesh and then on the far object.

_____

_____

Exercise F

The ability of the eye to compensate for near and far vision is called **accommodation.** At
your next laboratory period be prepared to describe exactly what happens to the lens of the
eye when accommodation occurs. In question 11, no instructions having been given, there
was probably much variation in the descriptions because some chose to focus on the mesh and
some on the objects seen *through* the mesh. This demonstrates that even the phenomenon of
accommodation may be a source of variation in perception.

Figure 22-1 consists of eighteen small drawings. On a separate piece of paper describe
each picture as accurately as you can. *Do not consult* with your neighbor. This should be done
completely independently. When you finish your instructor will ask for your descriptions and
will discuss them. **Do not read on**—begin your descriptions now.

14. Are all the descriptions in the class identical? _____ Why or why not? _____

_____

15. Can you generalize the kinds of answers into two basic types? _____

What are they? _____ and _____

Your instructor will discuss the implications of your responses with you.

Summary

An attempt has been made, during this laboratory period, to explain the nature of perception and to point out some of the problems relating to the formation of percepts which are
relatively accurate representations of reality. We have pointed out that the stimulus, the sense
organs, and the mind can operate in such a manner as to distort the percept (that is, cause the
percept to be different from the stimulus).

FIGURE 22–1

The most important safeguard to valid perception is a recognition of the factors which may interfere with accurate observation, *and a will to overcome them.* It is important to develop an *objective attitude,* that is, a constant awareness that previous experience, the emotions, and the sense organs often operate in a manner which distorts the percept. It has been shown, for instance, that when one sees a series of objects which have no particular pattern, the mind has a tendency to cause us to see these objects in a familiar pattern. If a picture is unfinished (like a picture of a dog that is complete except for one leg which is only partly drawn), we might not notice this flaw unless it was called to our attention. The mind has a tendency to complete that which is almost complete so as to make the unfamiliar familiar.

To recapitulate: *The key to a valid relationship with reality is a desire for objectivity and an awareness of the pitfalls of the perceptual process.*

PART II. THE EYE—A REPRESENTATIVE SENSE ORGAN

It will be helpful for you to relate the abstractions comprising Part I of this laboratory exercise to the anatomy of a sense organ, in order to better understand the nature of sensation. In your examination of the eye, as required below, keep in mind the nature of perception and try to relate each functional part of the eye to its role in the perceptual process.

The Anatomy of the Eye

Each pair of students will be given a preserved sheep eyeball. Place the eyeball with the iris and pupil facing up, and identify the external parts as shown in Figure 22-2.

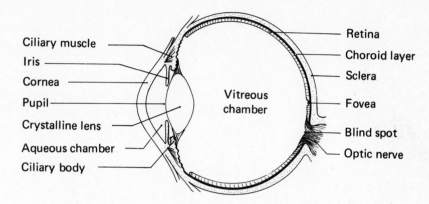

FIGURE 22-2   Sagittal section through a mammalian eyeball.

After you have finished the external morphology, cut the eyeball into halves by making a vertical incision through the pupil. Continue the incision completely through the eye. Place the two halves in a dish of water, with the inside of each half facing upward. Identify each of the structures on the inside of your sheep eyeball, as depicted in Figure 22-2.

Consult Figure 22-3, which represents the path of light as it enters the eye. Note that the cornea and the fluid-filled **aqueous chamber** bend (refract) the light first, after which it passes through the **pupil** of the eye. Figure 22-3 also shows how light rays striking the **iris** are prevented from entering the inside of the eye. As the light source gets brighter, fewer light rays are required to cause an image to be produced and the pupil is reduced in size. In darkness, the pupil is dilated by the contraction of muscle fibers in the iris, allowing more light rays to enter.

The light then enters the lens of the eye, where it is bent according to the lens shape, which is regulated by the contraction and relaxation of the **ciliary muscles**. A short squat lens will bend the light sharply, causing the rays to converge (focus) relatively close to the lens. A

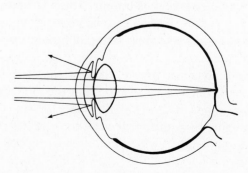

FIGURE 22-3   The refraction of light as it enters the eye.

long thin lens will cause the rays to focus far away from the lens. A lens which does not present a perfectly rounded surface (for example, having a bump or depression) will cause the image to have an astigmatic aberration or distortion of that portion of the image which is produced by the refraction of the distorted part of the lens (**astigmatism**).

The lens bends the light so that it focuses, ideally, on the **retina** of the eye in the region of optimal sensitivity, the **fovea**. The region of the fovea contains the greatest concentration of cones (one of the two basic types of light receptors of the eye) thus permitting maximal day-time vision. The retina, then, is the "screen" of the eye, upon which the image is focused.

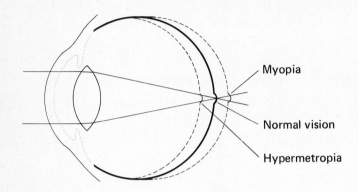

FIGURE 22-4   Myopia (nearsightedness) and hypermetropia (farsightedness). Note that in hypermetropia the eyeball is shortened while in myopia the eyeball is elongated.

Figure 22-4 shows two common eye defects, **myopia**, or nearsightedness, and **hypermetropia**, or farsightedness.

16. What happens to the image when each of these conditions occurs? _____

_____ Why? (Explain clearly for each condition.) _____

_____

_____

_____

The association neurons, which pick up the nerve impulses produced by the receptors of the retina, leave the eye at one point to form the **optic nerve**. This thick white tube can be seen leaving the posterior of the eyeball. The optic nerve carries the impulses to the brain, where they are interpreted as visual images.

The region of the retina which is just anterior to the optic nerve is made up of outgoing neurons and does not contain receptors. It is not receptive to light and is called the **blind spot**.

The **aqueous chamber** contains a watery fluid called **aqueous humor** which is optically more dense than air and thus makes the cornea a kind of water-filled lens. The **vitreous chamber** is filled with a transparent jelly-like fluid which transmits the light after it passes through the lens.

The walls of the eyeball are composed of three layers of tissue. The outermost layer is the **sclera**. It is thick and tough and protects the eye. It becomes transparent at the anterior of the eyeball to form the cornea. The second (middle) layer of the eyeball is the **choroid layer**. This layer is darkly pigmented and absorbs excess light. In addition, it is richly supplied with blood vessels. The **retina** is the thin flesh-colored inner layer which contains the cones and rods, the receptors of the eye.

Summary

17. List the parts of the nervous system, including the parts of the eye, employed in the formation of a visual percept. Next to those parts, where applicable, indicate how the percept can be modified or distorted so that each image is a personal reflection of reality. Start with the stimulus and work your way inward.

Example
1. Cornea: The curvature of the cornea and its transparency can modify the image by blurring or distorting it, as in astigmatism.

VII Bacteria and Disease

23 Nature of Proof—Koch's Postulates

Causative Agents of Disease: "Sick" Carrots *

Most of us have been afflicted with childhood diseases. Though sometimes serious, diseases like measles and mumps are considered to be relatively harmless accompaniments of childhood. But more serious children's diseases which afflicted your parents and grandparents, such as whooping cough, scarlet fever, and tuberculosis, are now rare. The infant death rate 50 years ago was 9.2%; now it is 1.9%. Less than 150 years ago the childhood death rate was many times higher, for no one understood the nature of infection, and bacteria were only suspected—not proven—to be the causative agents of disease.

A controversy concerning the role of bacteria in disease raged throughout the first half of the nineteenth century, and as late as the 1860s Joseph Lister was unsuccessful in his efforts to convince midwives and physicians who assisted in childbirth to wash their hands. Millions of women were dying of "childbed fever" because their physicians went from birth to birth unwashed, spreading the bacteria which caused the disease.

Progress in the treatment of disease was halted, awaiting the final proof that bacteria cause disease. Robert Koch, in 1884, provided that proof. His experiments have become classics, and are important to us because they reflect on the nature of proof in general. All of us face situations where it becomes necessary to *prove something to be true*. The most obvious example is in a court of law. The prosecutor must prove that the defendant has committed a crime. Undoubtedly you will some day be on a jury of citizens responsible for evaluating the validity of his proof.

Today we will attempt to follow Koch's reasoning and develop his system of proof for ourselves.

PRELIMINARY INFORMATION

At first glance the problem of proving that bacteria cause disease seems simple—perhaps Louis Pasteur thought so when he identified bacteria in the blood of diseased sheep. Pasteur found rod-shaped bacteria in the blood of sheep which had the disease anthrax. He suggested

---

*The idea for this exercise was provided by Daniel D. Burke in Jon C. Glase, Ed., *Tested Studies for Laboratory Teaching*, Kendall/Hunt, Dubuque, Iowa, 1980.

that the bacteria caused the disease. There were numerous criticisms of the logic behind this conclusion. Analyze the following description of Pasteur's work as a basis for understanding the potential for this criticism.

Read the paragraph below, then place a check next to each of the statements (A–R) following it that you believe to be true.*

(1) Louis Pasteur was interested in the disease of animals and man called anthrax. (2) He had previously demonstrated that sheep can get the bacterium *Bacillus anthracis*, from the soil in which it can live indefinitely. (3) He wondered how the bacillus got into the soil. (4) At first he thought that the bacteria simply crawled around until they found a host they liked. (5) In inspecting farms where many sheep had died of anthrax, he noticed certain dark, wet areas of the surface soil. (6) Farmers told him those were the sites where they had buried sheep that had died from anthrax. (7) Careful inspection revealed numerous worm castings on the surface of these dark, wet spots. They were more numerous than on the dry areas. (8) He thought that perhaps the worms ingested material from the diseased carcass and brought the germs to the surface where they infected the grass. (9) He cultured the castings on dry soil and also on the wet burial areas. (10) He found bacteria in the latter. (11) He crushed worms from the burial area and injected the fluids into sheep. (12) Many of these sheep soon died of anthrax. (13) He decided that the worms brought anthrax bacilli to the surface. (14) He suggested this might occur in the transmission of other animal diseases. (15) He urged that animals that died of infection should not be buried in pasture land.

Place a check next to each statement below you believe to be true.
A. Statement (2) is a fact.
B. Statement (3) is a hypothesis.
C. Statement (4) is anthropomorphic.
D. Statement (3) is the statement of the problem.
E. Statement (7) is an observation.
F. Statement (8) is an assumption.
G. Statements (2) and (1) are part of an experiment.
H. Statement (11) is part of an experiment, but not statements (9) and (1).
I. Statement (12) is teleological.
J. Statement (12) is a conclusion.
K. Statement (13) is a deduction.
L. Statement (13) is an induction.
M. Statement (13) must be correct, for the preceding experiment [statements (11) and (12)] proves it to be correct without a shadow of a doubt.
N. It is possible that the crushed worms in statement (11) could have caused the disease themselves, and the bacteria were just artifacts (incidental).
O. Pasteur showed that fluids squeezed from worms [statement (11)] caused anthrax, but he didn't prove that any specific bacteria in the fluids caused the disease.
P. The fact that more bacteria were found in the wet burial area [statement (10)] than in the dry area means that the bacteria were anthrax-causing bacteria.
Q. Many of the bacteria found in the burial area [statement (10)] may have been feeding on the buried sheep and may not have been *Bacillus anthracis*, the presumed cause of anthrax.
R. The paragraph, taken as a whole, makes clear beyond doubt that Pasteur proved that *Bacillus anthracis* causes anthrax.

*This material is taken from a booklet of tests. Its author is unknown.

The requirement for proof that underlies science caused a number of other scientists to test the hypothesis that bacteria cause the disease anthrax. Some scientists inoculated the blood of diseased animals into healthy sheep, which promptly became sick—but no bacteria were found in their blood. Other workers used blood from sick sheep in which *no bacteria were visible*. The sheep receiving the blood contracted anthrax and *rod-shaped bacteria were found in their blood*. Other experiments showed that anthrax was somehow associated with certain places—that is, something in the soil transmitted the disease. Cattle and sheep kept in these areas came down with the disease while those kept away from them remained healthy. Thus, it seemed, bacteria in the blood had nothing to do with the disease. Perhaps the rod-shaped structures in the blood of diseased animals were *symptoms* of the disease, not the causative agents. Perhaps they were completely incidental, their presence or absence having nothing to do with the disease.

It had become necessary to develop a system to prove beyond a shadow of doubt that a proposed agent was truly the cause of a disease. During today's laboratory period you will have a chance to develop such a system as you attempt to prove the presence of a causative agent which makes carrots "sick."

■ THE PROBLEM: *To identify the cause of disease in sick carrots and to prove our hypothetical cause to be correct.*

PROCEDURE

Examine the pair of petri dishes in front of you, marked 1 and 2. Open each dish; use a wooden probe or a scalpel to poke each slice of carrot. Examine the slices carefully—you might even want to smell each slice. One of the slices is firm and crisp. The other, cut from an ordinary carrot at the same time, is "sick."

1. Describe the symptoms of the disease. _____

_____

_____

_____

You will be given slices of fresh carrot and petri dishes. Bunsen burners are to be used to flame the tip of your scalpel to remove any live microorganisms which might be disease-causing.

Procedure for Transfer of Microorganisms

(1) Place a closed petri dish containing four carrot slices in front of you.
(2) On a piece of paper toweling place a pair of forceps (tweezers), a scalpel, two sterile disposable petri dishes, two clean microscope slides, and a clean bacteriological inoculating loop.
(3) Obtain a squeeze bottle of sterile distilled water and an oil immersion microscope.
(4) Place a closed petri dish containing 70% alcohol to your right.
(5) Place your bunsen burner on your left, so that it is away from the petri dish with

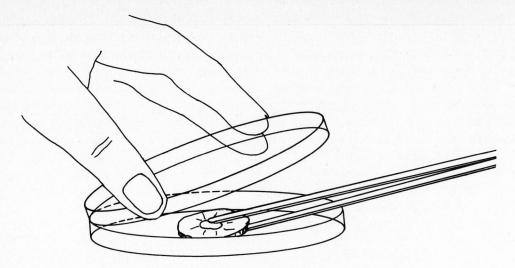

FIGURE 23-1 Inserting carrot slice into sterile petri dish.

the alcohol in it. Light the bunsen burner and adjust it so that the flame contains an inner blue cone.

Caution: Avoid accidentally lighting the alcohol fumes by keeping the petri dish closed and away from the flame.

(6) Pick up a slice of fresh carrot with your forceps. Dip it into the dish of alcohol for a moment and then touch the carrot to the bunsen flame. It will catch fire briefly. Place the carrot slice into a sterile petri dish by tilting up the cover of the dish and sliding the carrot slice inside (see Figure 23-1). Close the lid and repeat the procedure with another slice of carrot and the other petri dish.

(7) Hold the scalpel in the blue part of the bunsen flame for about 10 seconds. Allow the blade to cool for several moments.

(8) Tilt up the cover of the petri dish containing the sick carrot (making sure the cover is still above the bottom plate to prevent dust from contaminating the dish). Insert your sterile scalpel and remove a small piece of sick carrot tissue.

(9) Transfer the sick carrot tissue to a slice of flamed fresh carrot in one of your two petri dishes. Smear or squash the tissue of the sick carrot on the surface of the experimental carrot slice (see Figure 23-2).

(10) Add 5 ml of sterile distilled water and replace the lid. Add 5 ml of sterile distilled water to your other (control) petri dish.

(11) Place your initials on the petri dishes and put the dishes on a tray at the front of the room. The tray will be placed in an incubator.

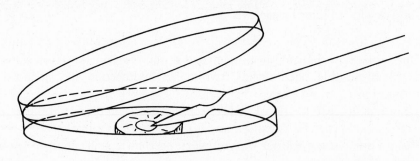

FIGURE 23-2 Smearing infected carrot tissue on surface of experimental carrot slice.

This technique will allow you to test a hypothesis. But as yet you have not had a chance to think about the problem. Use the space below to state a hypothesis and to write a protocol or plan to prove your hypothesis correct. Remember, you are trying to prove that some suspected causative agent is making your carrot sick.

☐ Hypothesis:_____

_____   _____

Protocol

Obtain a prepared microscope slide labeled with a number from 1 to 24. Each slide was made from a different sick carrot, but each sick carrot had the same symptoms. Examine your slide under the oil immersion lens of your microscope. Your instructor will demonstrate the use of the oil immersion lens.

Caution: Do not attempt to use the oil immersion lens until its use has been demonstrated by your instructor.

Begin by focusing on the blue-stained film under low power. Find an area that looks like a mosaic of crossed lines, with spaces between the lines. Switch to the high dry (40X) lens. Focus carefully with the fine adjustment. Move the high power lens out of the way, place a drop of immersion oil in the circle of light, and swing the oil immersion (100X) lens into place. Focus carefully with the fine adjustment. The lines you see are masses of bacterial. *Look in the clear spaces for individual bacteria.*

Record your observations._____

_____   _____

When each member of the class is finished, the instructor will summarize the observations on the chalkboard.

Record the class's observations._____

_____   _____

2. What is the significance of these observations? (Remember, each slice was taken from a different sick carrot, but all carrots exhibited the same symptoms.) _____

_____

3. Do they prove conclusively that the bacteria are the causative agents for the disease?

_____Explain why not._____

_____

_____

4. Do they suggest that the bacteria might be the causative agents? _____

I. State precisely the information you have just learned concerning sick carrots. _____

_____

_____

(Refer to this later—related to Koch's postulate 1.)

We now know that sick carrots have rod-shaped bodies in fluid taken from their tissues, but this does not constitute absolute proof that the rod-shaped bodies are causing the disease.

5. What must we now do to prove that the *rod-shaped objects* are the disease-causing

agents? _____

_____

_____

You will find a test tube containing sterile solid medium at your table. The medium supports the growth of bacteria. A week ago the medium was inoculated with the rod-shaped objects *taken from sick carrots*. Note that a thin whitish film coats the surface of the medium, indicating bacterial growth. Test tube B, the control, was not inoculated with bacteria. Notice that its surface is clear.

Flame the tip of a bacteriological loop until it glows and let it cool for a few moments. Tilt test tube A and remove the cotton plug (Figure 23–3a). Flame the mouth of the test tube (Figure 23–3b). *Gently* scrape the surface of the medium, removing some of the whitish material, and smear it on a clean slide. Add a drop of sterile distilled water and mix the two with the tip of the loop. Allow to dry. Pick up the slide with your forceps and pass it through the bunsen flame several times. This will fix (coagulate) the protein of the bacteria and, at the same time, make the preparation adhere to the glass. Add 1 drop of methylene blue stain. Allow to stand for 1 minute. Wash off with a gently flowing stream of tap water or dip into a beaker of water and allow slide to air dry. Examine under the oil immersion lens of your microscope.

What are your observations? _____

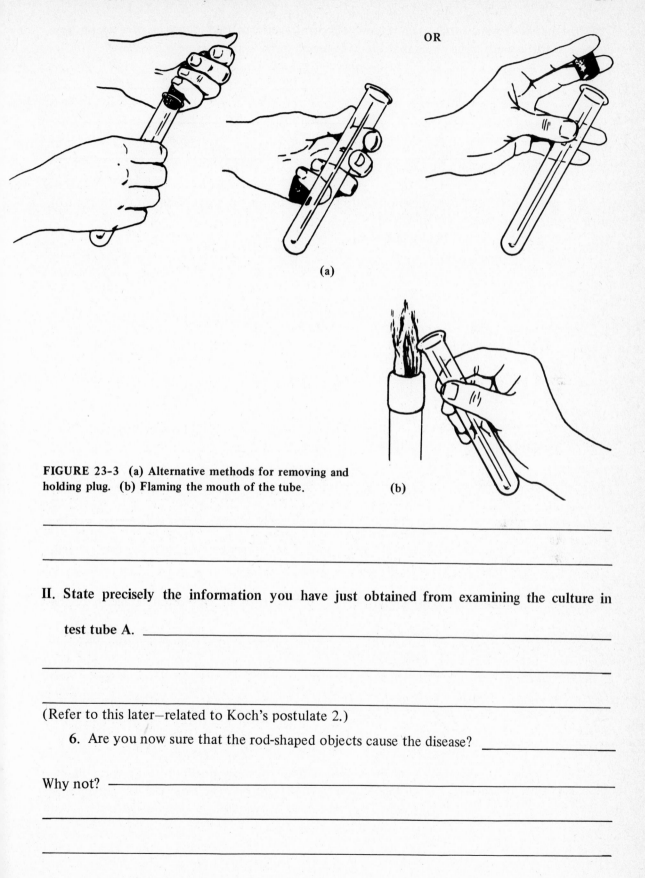

FIGURE 23-3 (a) Alternative methods for removing and
holding plug. (b) Flaming the mouth of the tube.   (b)

_____

_____

II. State precisely the information you have just obtained from examining the culture in

test tube A. _____

_____

(Refer to this later—related to Koch's postulate 2.)

   6. Are you now sure that the rod-shaped objects cause the disease? _____

Why not? _____

_____

_____

III. Now that we have a pure, uncontaminated culture of the suspected causative agent, how can we test further our hypothesis that the rod-shaped objects are causing the disease?

_____

_____

_____

(Refer to this later—related to Koch's postulate 3.)

Use petri dishes and two slices of healthy carrot to set up an experiment, with control, which would further our quest for proof.

Describe what you will be doing below.

7. What do you *expect* will happen? _____

_____

_____

Suppose that your expectations are valid. (You will determine their validity next period, after bacteria have had a chance to reproduce in huge numbers.)

8. Is this the final proof that the rod-shaped organisms are the causative agents of the

disease? _____ Explain at length. _____

_____

_____

_____

Koch, goaded by his critics and unsatisfied if there was any possibility of error, added a fourth and final postulate: *The suspected bacteria can be isolated from the experimentally infected host, cultured, and shown to be the same as the bacteria in the original culture.*

We have now followed the four procedures that were evolved by Koch to prove that bacteria cause disease. Other scientists, using his technique, have traced the causes of most bacteria-induced diseases, providing the opportunity to combat these diseases. Thus the number of child-killing diseases has been reduced drastically, and the length of the human lifespan enlarged dramatically.

Analyze the procedures you followed today and record below the four steps of the proof. This is the most important part of this exercise, since it requires you to put together all of the hitherto seemingly unrelated procedures you used.

Step 1

Step 2

Step 3

Step 4

REPORT SHEET   Name _____

Instructor's name _____

Record the results of your experiments after an incubation period of at least 48 hours. You will be allowed a few minutes at the beginning of next period to examine your petri dishes. Your instructor will show you where to dispose of your experimental petri dishes.

	Experimental Slice	*Control Slice*
1. A piece of sick carrot was rubbed on a slice of healthy carrot.		
2. A pure culture of bacteria from a sick carrot was placed on a slice of healthy carrot.		

Do you feel that you have proved the cause of the sick carrots? _____
Explain in depth (write a short essay summarizing the relevant portions of this exercise).

VIII Mechanisms of Inheritance

24 Synthesis of Simple Observations into Complex Concepts

Meiosis

In the latter part of the nineteenth century, Edouard van Beneden, a Belgian parasitologist-physician, examined the ovary and uterus of a species of roundworm, *Ascaris lumbricoides*. This worm produces hundreds of eggs which gradually mature as they pass down the uterus toward the vagina. As the eggs leave the ovary they are penetrated by bullet-shaped, tailless sperms stored in the **seminal receptacle**, a chamber located next to the ovary. (The sperms were previously placed in the lower reaches of the reproductive tract by the male and have propelled themselves up the uterus by ameboid movement.) Since the eggs are penetrated as they leave the ovary, it follows that those eggs found in the lower portion of the uterus have contained sperms longer and are more mature than the eggs nearer the ovary.

In large measure the preparation of the nucleus of *Ascaris* eggs for fertilization occurs *after* the sperm penetrates the egg, but before the nucleus of the sperm reaches the nucleus of the egg. (In humans, some changes in the nucleus of the egg occur *before* sperm penetration, as the egg passes down the oviducts.) The fusion of the sperm with the egg is called **sperm penetration**; the fusion of the **sperm nucleus** with the **egg nucleus** is called **fertilization**.

Because the eggs of *Ascaris* are transparent, and because their preparation for fertilization takes place after the sperm has penetrated the egg, van Beneden was able to observe the changes in the nuclei of the egg and sperm which preceded fertilization. A representation of what he saw is depicted in Figure 24–1.

Although his original interest in the reproductive organs of *Ascaris* was not oriented toward understanding the nature of fertilization, the progression in the development of the nuclei of the eggs after the sperms had entered them fascinated him. Because the scope of his thinking was broad enough to permit him to appreciate the panorama unfolding before his

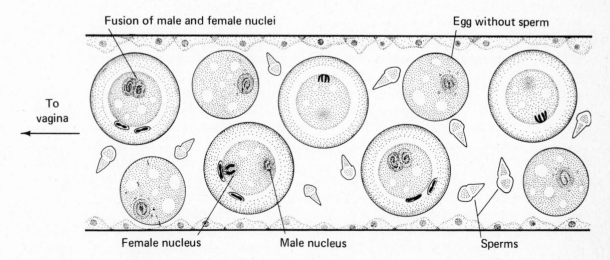

FIGURE 24–1 Uterus of female *Ascaris megalocephala* containing sperms previously introduced by a male.

eyes, van Beneden was able to describe for the first time a key step in the process of reproduction, the halving of the nuclear material of the sperm and egg.

It should be remembered that the mechanism for heredity had not yet been clearly deciphered when van Beneden made his discovery, and the role of the nucleus in heredity had not yet been elucidated. Van Beneden's contribution can be summarized simply:

> *The nuclear material in the egg and sperm is reduced by one half before fertilization occurs.*

The implications of this simple statement are profound. Once it was established that the nuclei are reduced to one half their normal mass in the sex cells, this suggested that in them lies the vehicle for the transmission of hereditary characteristics. Only the halving of the nuclei can account for the fact that all the cells of the offspring resulting from the fusion of the egg and sperm have the normal amount of nuclear and cellular material. Previous to van Beneden's discovery there were some who believed that the sperm brought a "vital force" to the egg, infusing life into the latent egg and stimulating it to develop. Others believed they saw little humans (homunculi) inside human eggs or sperms, and that these homunculi, once stimulated to develop by the fusion of the egg and sperm, simply enlarged to become the offspring. Darwin believed in **pangenesis**, the development of minute particles inside the egg and sperm into the organs of the body, each particle being predestined to become a particular organ. (To put this in proper perspective, remember that all of this thinking occurred less than 150 years ago.)

Concepts based on the transmission of a "vital force" or other supernatural or otherwise inexplicable factors are termed **vitalistic**. They are representative of a school of biological thought which maintains that there is an intangible factor found in all living things which, by its immaterial nature, is not susceptible to study by science, science being based on **empirical fact**.

Those arguing against this idea are called **mechanists**. They maintain that "life" can be reduced to a finite number of physical and chemical processes, and that it is conceivable that some day man will learn about these processes and be able to create life in the test tube.

Since we have not yet been able to manufacture "life," it cannot be said that either school is absolutely correct.

It can be seen, however, that the mechanistic point of view is much more fruitful for the scientist since it presents a challenge to science to prove its correctness; that is, it suggests that experiments can be employed to test the validity of the argument.

Van Beneden showed a simple physical reduction in the mass of the nucleus, and by so doing suggested that a mechanistic or purely physical solution to the problem might be possible. Once scientists were oriented toward a mechanistic answer to the question of the process of heredity, it became possible to understand the role of the chromosomes.

In the following laboratory exercises on meiosis and heredity, you will be tracing some of the reasoning of the scientists who solved the riddle of heredity. You will be presented with the same kinds of problems and you will be required to arrive at solutions to these problems. You have a distinct advantage over your illustrious predecessors, however. You already know that chromosomes and genes are the vehicles for the transmission of hereditary characteristics.

As you attempt to solve each problem in this exercise, carefully relate your decisions to the information you have obtained through the solution of previous problems. In this manner you will be duplicating the pyramidal development of any scientific concept, each new idea being suggested by several previously worked-out concepts. If you have solved a problem incorrectly, the answer will not fit into the rest of the data presented in subsequent problems. It will then be necessary to go back to alter the incorrect answer.

PRELIMINARY INFORMATION

Ascaris megalocephala is a roundworm (superphylum Aschelminthes, phylum Nematoda) which is an internal parasite of horses. A close relative, *Ascaris lumbricoides*, inhabits the small intestines of millions of human beings. If present in small numbers these worms do not cause severe symptoms; however, infection with great numbers may block the lumen of the intestines, preventing the proper functioning of the digestive tract.

Ascaris reproduces sexually; that is, two parents must contribute the genetic material which, when fused together at fertilization, forms the new organism. At your table will be found examples of male and female *Ascaris.* Note the hooked tail of the male.

Reproduction is initiated when the male places millions of bullet-shaped spermatozoa into the vagina of the female. The sperms eventually end up in the tubular oviduct and uterus, where hundreds of eggs have accumulated. As soon as a sperm enters an egg, a **fertilization membrane** forms around the egg, preventing the entrance of other sperms.

Ascaris megalocephala has all its characteristics determined by four (two pairs of) thread-like structures known as **chromosomes**.

1. If the parents have four chromosomes in each of their body cells, how many chromosomes would an offspring have in each of its body cells? _____

2. How many chromosomes would the third generation have? _____

3. The tenth generation? _____

If you have been doubling the number of chromosomes with each new generation, consider the fact that a condition known as **polyploidy** (where the daughter organisms have more sets of chromosomes than their parents) results in unnatural giant offspring. If you have reduced the number of chromosomes in the offspring to less than the normal number, sterility will result. Change your answers to questions **1, 2,** and **3** if they are incorrect.

4. If each parent has four chromosomes in its cells and each daughter organism must have a similar number, how many chromosomes should be found in each **gamete** (egg or sperm)

of the parents, if the total number resulting when they fuse is to be four? _____

5. If you have hypothesized that the number of chromosomes in each gamete is less than that in the fertilized egg (**zygote**), does this mean that the new organism will have fewer characteristics than the parent? _____ Explain. _____

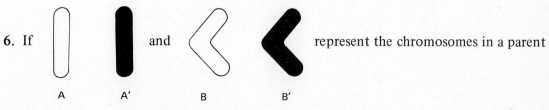

6. If   and   represent the chromosomes in a parent

    A   A'   B   B'

cell and each parent is able to give only half its chromosomes to the offspring, would normal offspring result if the parents contributed chromosomes A and A' (and thus did not contribute

B and B' chromosomes)? _____ (Assume that A and A' are similar in their effect and different from the effect of B and B'.)

7. What might happen to the appearance of the offspring if the contribution of chromosomes was a completely chance phenomenon (so that the situation outlined above could

actually occur)? _____

_____

8. If a male had A, A', B, and B' chromosomes, what possible alternative arrangements

of chromosomes could be found in his sperms? _____ _____ _____

_____ (Four combinations of two letters each.)

9. Suppose that the chromosomes in the egg were not united with the chromosomes of a sperm, yet a stimulus caused the egg to develop into an organism anyway. (This occurs both naturally and artificially and is called **parthenogenesis.**) *The resultant organism appears to be physically normal.* Explain. (Keep in mind the conclusions you have previously drawn from the

fact that at least one chromosome from every pair is contributed to the offspring.) _____

_____

_____

10. Certain insects such as bees normally produce several kinds of eggs, one type being unfertilized. This unfertilized egg develops into a normal-appearing bee. What inference must you make about the effect of individual chromosomes (one member of each pair of chromo-

somes)? (Question 9)* _____

_____

11. Did the bee have the same number of chromosomes in its cells as are found in the

cells of a bee formed by the union of a sperm and egg? _____

12. What percentage of the total number did it have? _____

13. A frog egg will also develop parthenogenetically if pricked with a needle dipped in

salty water. What must be the parthenogenetic frog's sex? _____ (In frogs and in humans, maleness is controlled by a chromosome contributed by the father.)

14. Why is it unlikely that a sperm would develop parthenogenetically, that is, without

*If you are not sure of your answer, refer to the question in parentheses for more information.

the benefit of the egg? (Consider the physical characteristics of each—size and composition—

in formulating an answer.) _____

_____

_____

Let us determine how the egg and sperm are formed so that they contain half of the original number of chromosomes. Certain facts will be presented to help you formulate your hypothesis.

Section A. A chromosome begins its existence as one very long DNA molecule. This molecule, some time before the actual process of division, replicates into two identical DNA molecules and forms two intertwined strands called chromatids. Together these form a chromosome which is capable of entering into division.

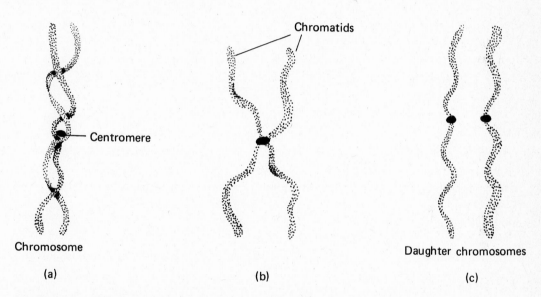

FIGURE 24–2 Chromosome (a) separating into chromatids (b) and daughter chromosomes (c).

Definitions (Figure 24-2)

Chromosome: A thread-like structure found in the nucleus of the cell which contains the blueprint for the hereditary characteristics of the organism.

Chromatid: Each partner of the pair of threads constituting the chromosome.

Daughter chromosome: When the chromatids separate, each is called a daughter chromosome.

Section B. The process of freeing the daughter chromosomes and distributing them to the sperms and eggs is called **meiosis**. During this process there are two divisions. The first reduces the original number of chromosomes (four in *Ascaris*) by one half by distributing the chromosomes between two daughter cells. After the first meiotic division, then, there are two cells, each containing half the number of chromosomes found in the original cell (in *Ascaris* two daughter cells each containing two chromosomes).

The second meiotic division is accompanied by a separation of the two chromatids which make up each chromosome into two daughter chromosomes. The two daughter cells divide, so that the end result of meiosis is four cells. The chromatids separate (and are now called daughter chromosomes). They are distributed equally among the four cells. In *Ascaris,* from the original cell with four chromosomes, we have four daughter cells each with two *daughter chromosomes.*

In *Ascaris*, then, the sperms and eggs contain two daughter chromosomes each, and their fusion will result in a fertilized egg with four *daughter chromosomes.* Eventually each single-stranded chromosome will reduplicate itself, and by the time it is ready for cleavage the fertilized egg will contain four double-stranded chromosomes.

15. If you end up with two daughter chromosomes in each sperm or egg, and you start

off with four chromosomes (or _____ potential daughter chromosomes), what proportion

of the original number of daughter chromosomes is present in the egg or sperm?_____

16. How many divisions must have occurred? _____ (Check with the second sentence in section B to see if you are correct.)

17. How many eggs or sperms could result, then, from the two successive meiotic divi-

sions of a cell with four chromosomes? _____ (Check with last sentence in section *B*, paragraph 2.)

Production of gametes begins with the cells lining the ovaries and the tubules of the testes. These cells are called **primordial germ cells** and are similar in appearance and chromosome number to the epithelial cells lining the tubes and chambers of the body. The first step in the formation of the sperm or egg is the unwinding of the chromosomes to form chromatids.

18. In *Ascaris,* before any division occurs, how many chromatids would there be (*Ascaris*

has four chromosomes)? _____ (Each pair of chromatids sticks together and moves as one unit in the first meiotic division.)

19. Draw the unwound chromatids still attached to one another in the *Ascaris* germ cell.

Remember, the chromosome number is two pairs, or four, and would look like this:

Depict the paired chromatids of each chromosome with a different color. *Use four colors,* so as to represent each of the four pairs of chromatids. The chromatids, since they are attached, should look like this: ❯❯ If they were rewound back into a chromosome, they would look like this: ❯

Ascaris **germ cell**

20. Draw the first meiotic division in which two cells are formed. *The chromatids are still stuck together.*

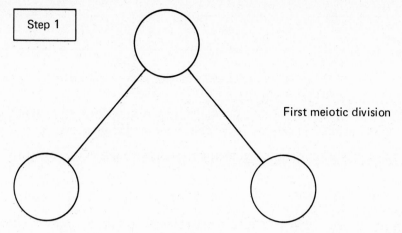

Step 1

First meiotic division

21. How many chromatids are there in *each* daughter cell of **step 1** of meiotic division?

_____

22. How many chromosomes would we have if we rewound the chromatids? _____

23. Have we reduced the original number? _____

Remember, however, that each chromatid is really a potential daughter chromosome and has the power to produce all the characteristics the original chromosome was able to produce.

24. Examine the original primordial germ cell. We represent each chromosome by a different color to indicate that each was capable of producing somewhat *different* characteristics in the offspring. Now examine the two cells in **step 1** of meiotic division which were derived from the primordial germ cell. Does each daughter cell contain the same kind of chromatids (or chromosomes) present in the original primordial germ cell? In other words, does each daughter cell contain exactly the same genetic message as the original cell from

which it was derived? _____ Describe two differences between the original cell and the daughter cells.

a. _____

b. _____

One of the differences described above should have led you to understand that the process of gamete production, even at the first meiotic division, provides for variability between the characteristics of the parent and those of the offspring.

25. Explain how variability (differences between parent and child) is provided for even

at the first reduction division. _____

_____

_____

26. Is variability desirable? _____ Why or why not? _____

_____

_____

Step 2 of meiotic division begins with one of the two daughter cells of the first meiotic division (one of the two cells from **step 1**).

In **step 2** the chromatids separate into daughter chromosomes and two daughter cells are formed from each of the cells in **step 1**. These cells become the gametes.

27. Sketch the daughter chromosomes in **step 2**

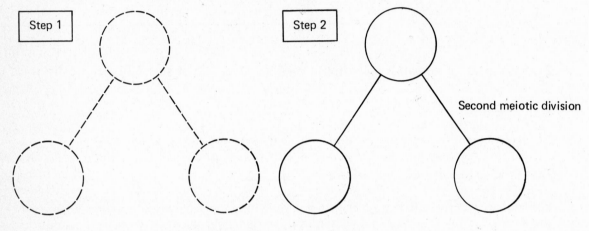

28. How many daughter chromosomes are there in each cell? _____

29. If two such cells fuse (for example, a sperm with an egg), what is the number of

daughter chromosomes in the nucleus of the resultant cell? _____

30. In the cell resulting from the fused gametes, each daughter chromosome eventually reduplicates itself and will consist of two chromatids joined together by the centromere. How many chromosomes (each consisting of two chromatids) would there eventually be in this cell

in *Ascaris*? _____

If your answer is not four, or the normal number in *Ascaris* cells, then start again working from both ends of the problem and continue until you see that the two successive meiotic divisions result in gametes with two daughter chromosomes each. (See especially question **19**.)

31. How many sperms can be produced by one primordial germ cell? (Question **17**) _____

In most cases of sperm production, however, the primordial germ cell undergoes an ordinary nonreduction **mitotic division** once or twice before beginning its meiotic division. Thus the process of gamete formation usually results in at least twice as many sperms as would be produced if the primordial germ cell participated directly in meiosis.

Egg production is the result of two meiotic divisions, paralleling sperm formation.

However, in humans only one egg is produced. This occurs because one of the cells ends up with virtually all the cytoplasm. The other cells formed in the meiotic division cannot function adequately, either in terms of forming food for the developing embryo or in the maintenance of the nucleus. These cells are called **polar bodies**. They eventually disintegrate.

32. How many polar bodies would you expect to be produced with every egg? (Question 17) ____ In some organisms, the first or primary polar body does not divide further. In this case you would have (number) _____ polar bodies produced with every egg.

33. If an organism has four chromosomes in its body cells, how many chromatids would it have in the primary polar body produced from **step 1**? (Question 21) _____ . How many daughter chromosomes would it have in the secondary polar body produced from **step 2**? (Question 28) _____

34. In summary, it is possible to have two slightly different modes of polar body production. One process exactly parallels sperm production, resulting in the formation of _____ polar bodies, each having the same number of daughter chromosomes. The other process has _____ polar bodies because the primary polar body does not divide.

Polar bodies appear as small dark bumps on the outside of the cytoplasm of the egg. The large circle in the following drawing represents the outline of an egg. Draw in the daughter chromosomes of the egg and polar bodies. This hypothetical organism produces three polar bodies and has four chromosomes in its body cells.

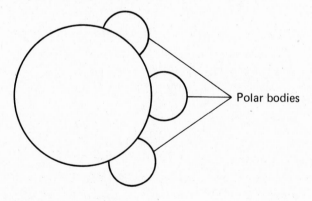

Reconsider the original chromosomes:

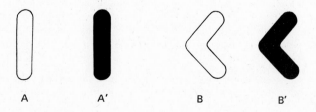

A   A'   B   B'

35. Draw the chromatids of a primordial germ cell, indicating A and B chromatids in white and A′ and B′ in black.

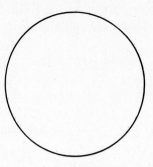

36. Indicate, by means of black and white chromatids, all possible combinations of chromatids which would appear in the daughter cells in **step 1**. (Members of each pair of chromatids remain attached to each other.)

37. Is it possible that as the chromosomes move about in the nucleus only dark chromosomes can end up in one nucleus, with the corresponding light chromosomes ending up in the

other nucleus? _____

38. Can there be a mixture of dark and light chromosomes in one nucleus?_____

39. Indicate by means of black and white daughter chromosomes, all possible combinations which would appear in the gametes (**step 2**).

Genes

By examination of the behavior of chromosomes it was found that the determinants of heredity were arranged in linear fashion on the chromosomes like beads on a string. Gradually more information supplementing this concept of hereditary units became available, and two collaborating scientists, one American and one British, won a Nobel Prize for their concept of the double-stranded helix containing deoxyribose nucleic acid (DNA).

Before this model was created, scientists needed a term to describe the then unknown heredity determiners found on the chromosome. They coined the term **gene** to refer to these factors. This term did not refer to any real structure, and even now it has a certain vagueness, being used to label various sequences of molecules comprising the DNA.

A gene is that portion of the chromosome responsible for determining some aspect of the organism containing it. Most characteristics are determined by the interaction of many genes. We will represent a gene by a stripe on a particular part of a chromosome, and we will sometimes refer to one gene as the determinant of major traits such as body size or cuticle color. This is a simplification used for teaching purposes; it does not mean that a complex trait like body size is determined by the action of only one pair of genes.

Consider Figure 24–3, showing the positions of genes for cuticle (skin) color and for body size. (Remember, most traits are determined by an interaction of many genes. For simplicity of explanation we will consider each trait below as though it were determined by one set of genes.)

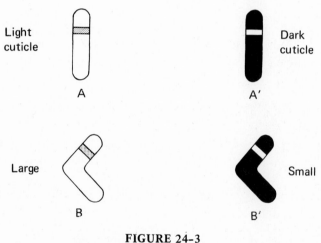

FIGURE 24-3

40. Take a cell with chromosomes A, A′, B, and B′ in it and draw the chromatids of the primordial germ cell, depicting the genes which are found on them. For example, chromosome A would yield chromatids like this:

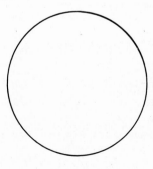

41. Draw below all possible combinations of chromatids appearing in **step 1**. (Remember, the chromatids are still stuck together.)

42. Is there any possibility that, even at this stage, provision for variation from the

appearance and genetic makeup of the parent organisms can occur? **(Question 24)** _____

43. What possible combinations of genes for cuticle color and for body size can come together in the cells of **step 1**? (Use words such as light cuticle, large, dark cuticle, small.)

_____

_____

44. Draw the four kinds of cells of **step 2**. Below each drawing write the possible combinations of genes for cuticle color and for body size in each of them.

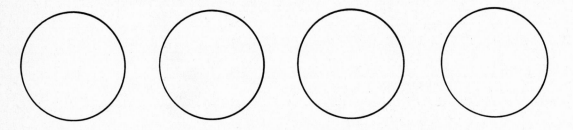

45. Assuming that both parents have the chromosomes A, A′, B, B′, is it possible for their offspring to have a different combination of chromosomes?

_____

_____

_____

(The four cells of **step 2** can be either four sperms for the male parent or one egg and three polar bodies* for the female.)

If we wanted to find out what possible kinds of offspring we could get from a male with a gene for a light cuticle (represent this as D), a gene for dark cuticle *(d)*, a gene for a large body

_____

*In the interest of accuracy, you are reminded that *Ascaris* produces only two polar bodies. It is easier to understand the process of meiosis if, in this simplified case, we consider the number of potential gametes identical in both males and females.

(S), and a gene for a small body (s) mated with a female with two genes for light cuticle (DD) and two genes for a small body (ss), we would perform the following steps:

46. Write down the genes of male _____ and female _____ .

47. The genes for cuticle color are on one set of chromosomes and the genes for body size are on another set. Each sperm or egg will have one daughter chromosome from each

set. Why? (Question 7) _____

_____

48. Figure out what possible combinations of genes can go into the eggs and sperms of this mating. (Remember that one gene of each pair goes into each gamete.)

Possible Combinations of DdSs   *Possible Combinations of DDss*

_____   _____   _____   _____   _____   _____

49. Explain why each line has only two letters on it. (Question 28) _____

_____

50. If the above combinations represent the kinds of eggs and sperms produced, what possible kinds of offspring could you have from their union? (Use letters to indicate combinations. Remember, each offspring will have two pairs or four chromosomes, one member of each pair being derived from the egg and the other from the sperm.)*

Kinds of Genes   *Physical Characteristics*
(letters)   *(appearance)*

_____   _____

_____   _____

_____   _____

_____   _____

LABORATORY EXERCISE

You have been given the opportunity to build up an understanding of the process of meiosis and an awareness of its importance in heredity. Now you will have a chance to apply your new-found knowledge to an examination of the same kind of *Ascaris* eggs which led van Beneden to make his important discovery.

Obtain a slide entitled "*Ascaris megalocephala* Maturation" and hold it up to the light. Notice three thick grayish lines under the cover slip. These are sections of the oviduct and uterus of a female worm. Since the eggs are fertilized on leaving the ovary by sperms stored in

*The capital letter denotes that the trait produced by that gene is dominant. The corresponding small letter denotes a recessive gene. If dominant and recessive genes for the same trait are present, the appearance will be that of the dominant gene. Thus *Dd* would yield a light cuticle even though the gene for dark cuticle (*d*) is present.

an adjacent chamber (**seminal receptacle**), those that are closest to the **birth pore** (at the end of the uterus) will have been fertilized first and will have had the most time to mature. This area (the posterior portion of the uterus) is to be found on the lowest strip, right-hand corner. The top strip, left-hand portion, is the most anterior part. It is the upper oviduct and will contain eggs which have not yet been fertilized and sperms which have not yet entered eggs. The lowest strip will therefore contain the mature eggs which have polar bodies, while the uppermost strip will contain mostly unfertilized or just fertilized eggs.

Place the slide under the low power of the microscope so that the upper left-hand portion of the oviduct is under the lens. Adjust the image under low power and then switch to high. The small bullet-shaped dark bodies are sperms; you will probably not be able to see chromosomes in them.

51. How many chromosomes should there be in each sperm? _____

In this animal the egg nucleus will not undergo meiotic division until the sperm enters the cytoplasm and begins to move toward the egg nucleus. The sperm, when inside the egg, appears as a gray shadow.

52. As the sperm slowly moves across the cell, the egg nucleus begins to divide. At the same time, the metabolic rate of the egg increases drastically, and a **fertilization membrane** appears and is thrown off from the surface of the egg, prohibiting the entry of other sperms.

Why? _____

_____

_____

_____

At one stage of its formation, the **first polar body** contains two chromosomes whose chromatids are spread apart, giving the impression of four thin strands. The egg nucleus will also contain four thin strands, so that you will see two groups of four chromatids when you view this stage of the formation of the first polar body.

Another view of the first polar body shows a black dot on the surface of the fertilization membrane, usually projecting downward into the **perivitelline space** (see Figure 24-4). The **second polar body** often appears as a black dot on the surface of the egg, projecting upward into the perivitelline space.

Since the primary polar body does not divide, a fully matured egg will have only two polar bodies—the primary, a black dot projecting down from the fertilization membrane, and the secondary, a black dot projecting up from the surface of the egg proper.

■ THE PROBLEM: *By using the information you have previously induced, find the cells mentioned below and draw each.*

First draw a schematic sketch at the left side of the Report Sheet, showing what you hypothesize the cell will look like, and then find it.

When found, draw the cell next to your schematic sketch and label it. Make all drawings at least 2 inches high.

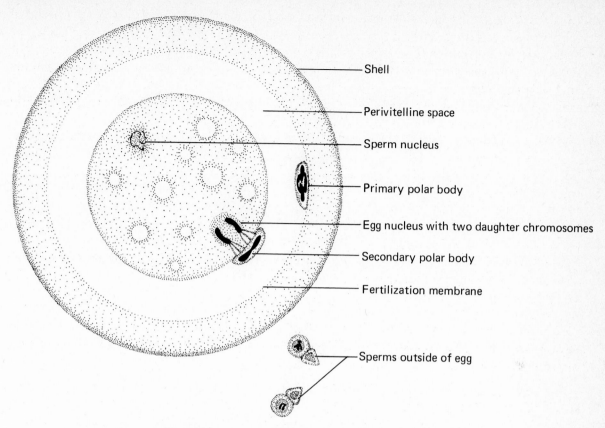

FIGURE 24–4 View of *Ascaris* egg containing a sperm.

The cells to be drawn on page 275 are

(1) An egg before sperm penetration. Show a sperm nearby. Note that the nucleus still has a nuclear membrane and that the chromosomes are not visible since they are extended in long, thin threads.
(2) An egg after first meiotic division. Should any polar bodies be present? If so, include them (it) in your drawing.
(3) An egg after second meiotic division. Include all polar bodies.

REPORT SHEET   Name _____

Instructor's name _____

Drawings of *Ascaris* Eggs

Schematic Sketch   *Drawing of Egg as Seen under Microscope*
(Draw before looking into microscope)

1. Before sperm penetration

2. After first meiotic division

3. After second meiotic division

25 Problem Solving in Genetics

It has been previously noted that in the early development of a science most of the subjects covered by that science were dealt with in a descriptive manner. Thus, chemists have spent much time describing elements and compounds, and much of biology has been descriptive in nature. Any science begins to come of age when a large part of its descriptive material is interrelated and explained by the establishment of theories, principles, and laws. The atomic theory, the major theoretical development in chemistry, has made possible the summarization, in terms of atoms, of large classes of facts.

In biology, the most significant theoretical concept developed thus far is the principle of evolution. It is the most important because there are more classes of facts interrelated and explained by this concept than by any other. The concept of evolution is not restricted to biology, however, but is successfully applied to nonliving things such as the earth and the universe.

The mechanism of evolution itself remained unexplained until the principles of heredity were worked out about 100 years ago by the Austrian monk Gregor Mendel, who, interestingly enough, had no concept of chromosomes. At the beginning of the twentieth century the "laws" Mendel proposed for inheritance, the visible effects of breeding, and the then recently discovered information about the behavior of chromosomes in meiosis all were joined to explain the phenomenon of heredity. Shortly thereafter, changes in heredity (mutations) were recognized as the basic raw materials of evolution. **The selection of mutations or groups of mutations by nature or man could explain the evolution of new species from old.**

The study of heredity is significant not only to the science of which it is a part but also to the practical world of affairs. The principles of heredity can be used in many practical plant- and animal-breeding procedures. They can be used to give a partial basis for understanding some social and personal problems such as race prejudice, some diseases, appearance of offspring, and determination of sex.

It ought to be obvious from the foregoing comments that the study of heredity is one of the most important subjects in biology. Because it is, and because many students are interested in this topic, considerable time is spent in its study.

This exercise is designed to give you an understanding of heredity. We will begin with the assumption that you know no more than that there is such a process as heredity and that it is somehow linked up with the reproductive process. If you think through the problems presented in the order of their presentation, without yielding to the temptation to pass over some of the more puzzling ones, you will have the satisfaction of arriving at a clear and logical understanding of heredity.

The procedure for attacking the problems in this laboratory exercise is as follows:

(1) Try to solve the problem yourself. Each solution requires knowledge built upon previously solved problems. When an answer is not clear to you, go back to the previous series of problems to clarify the theory. Then try again.

(2) To check your answer, find the number of the question in the right-hand column.

Do not look at the solution to a problem until you have made a vigorous attempt to find out the answer for yourself.

(3) If you have any questions after checking your answer, ask your instructor to help you. Do not consult with your neighbor; work independently.

SECTION A. THEORY OF MENDELIAN GENETICS

■ PROBLEM I

Compare the two family histories illustrated in Figure 25-1. The squares are used to designate males and the circles females. Horizontal lines connecting a square and a circle indicate a marriage. The vertical line leading downward from a horizontal line leads to another horizontal line from which other vertical lines lead to individual offspring. The Roman numerals indicate the generations. Note that albinism, indicated by the solid figures, appears in one family history but not in the other. An albino is a person having a deficiency of pigment in the skin, hair, and eyes.

IMPORTANT

The wide margin on the next pages will contain the numbered answers to the questions. Cover the answers with a book. Slide the book down the page and uncover each answer *after* you have written your own in pencil. Determine if your answer coincides with that in the margin. If not, go back to the question and try to find out why. If you cannot, ask your instructor for help.

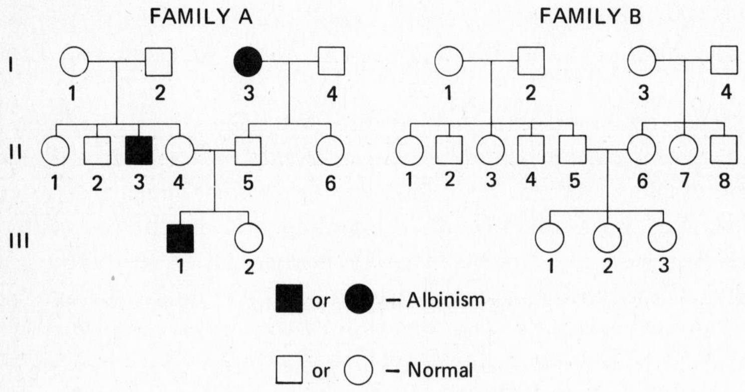

FIGURE 25-1 Pedigree charts of families A and B.

Compare the frequency of albinism in family A with that in family B. Also compare the frequency of albinism in family A with that in the population as a whole. The ratio of albinos to the entire population is approximately 1:200,000.

1. Does albinism occur more frequently in family A than in the entire population?_____

2. Since these two families lived in the same locality, the environment was the same for both. Therefore, ignoring the environment, how can you account for the fact that some members of the second and third generations in family A were albinos?

1. Yes.

The pedigree in Figure 25-2 is part of the history of family A. Study this pedigree and answer questions 3 through 25. Some of your answers will be hypotheses which you must think up for yourself on the basis of previous information. This will require considerable mental agility. Spend a good deal of time here. Be sure you get a reasonable answer to each of these questions before proceeding.

2. There must have been some hereditary factor passed on.

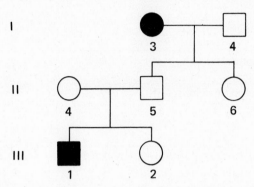

FIGURE 25-2 A portion of the pedigree chart of family A.

3. How many skin color conditions are involved?_____

What are they? _____

The term **characteristic** is defined as a specific inherited feature that can be observed. For example, blue eyes, curly hair, or the production of a specific enzyme are characteristics. Look at your partner or roommate and describe his or her characteristics for:

3. Two; albino and normal.

Hair color:_____

Eye color:_____

Left- or right-handedness:_____

The term **trait** is a general term to describe a set of characteristics. We use the general terms "eye color," "skull size," "hair texture," and so on, to refer to traits and the specific terms "blue eyes," "brown eyes," hazel eyes," and so on, to refer to characteristics.

In this family history, some individuals show a normal **characteristic** of the skin-color **trait** while others show the albino **characteristic** of the skin-color **trait**.

4. Study Figure 25-2. Which of the characteristics are present in

a. Generation I? _____

b. Generation II?_____

c. Generation III? _____

---

5. Did albinism disappear as an observable characteristic in

generation II?_____

4. a. Normal and albino.
   b. Normal.
   c. Normal and albino.

6. Has the characteristic reappeared in generation III?_____   **5.** Yes.

7. In the process of sexual reproduction what are the protoplasmic   **6.** Yes.

(cellular) contributions from parents to offspring called?_____

8. Does the individual in the four-celled stage of development have   **7.** Sperms and eggs (gametes).

hair, skin, and eyes through which skin pigmentation is expressed?_____

9. Would this stage of development show the albinism?_____   **8.** No.

10. What organs or structures must develop before the characteristic   **9.** No.

appears?_____

11. The environment does not cause albinism; the zygote (fertilized   **10.** Hair, skin, eyes.
egg) does not show the characteristic which eventually appears. On the
basis of the above facts and on the assumption that characteristics do
not arise without cause, do you infer that there must be something in
the zygote which caused the appearance of the albino characteristic in

the developing embryo?_____ We will call this "something" a **factor**.

12. Would you also infer that there must be a factor for normal   **11.** Yes.

pigmentation in the zygote which develops in a normal individual?_____

13. Where must the factor or factors in the zygote which caused the   **12.** Yes.

development of the characteristics have come from?_____
(Keep in mind the immediate origin of the zygote.)

14. Examine Figure 25-2 and note that individual III-1 is an albino.   **13.** Egg or sperm.
On the basis of this observable characteristic, what factor or factors

must this individual have?_____

15. On the basis of the appearance of the parents of III-1, what   **14.** Factor for lack of pigmentation
                                                                        (albinism).
factor or factors must each parent have?_____

16. On the basis of your answer to question 11, what can you infer   **15.** Factors for normal skin color;
                                                                        neither parent is albino.
concerning the origin of the factor or factors in individual III-1? _____

_____

17. On the basis of your last three answers, how many factors for   **16.** Completely hereditary.
pigmentation (that is, the trait albinism or normal) must parent II-5

have?_____What are the factors?_____

18. Study Figure 25-2. What evidence is presented in this figure to support your last inference? State a hypothesis to account for the fact that individual II-5 does not show albinism. (If you cannot make up what seems to be valid hypothesis here, ask your instructor for assistance.)

☐   Hypothesis:_____

_____

_____

19. Study family A of Figure 25-1. Do you find any reason for inferring that individual II-4 might carry a factor for albinism?

_____ If so, what is the reason?_____

_____

_____

20. How many factors for pigmentation does each individual in the pedigree have?_____(Remember that "pigmentation" refers to the amount of coloring matter in the skin. It may be normal in amount or it may be completely lacking, as in the albinos.)

21. Where does any one individual get the factors he possesses?

_____

_____

22. Make an inference to explain how each individual obtains the factors._____

_____

23. How many gametes are involved in the production of a new individual?_____ Assuming, on the basis of question **18**, that any one individual has only two factors for pigmentation, how many factors for pigmentation would you expect each gamete to carry? _____

24. To summarize some of the foregoing assumptions: Write the number of factors for the pigmentation trait in each parent.

_____ Write the number of factors for the pigmentation trait in each gamete. _____ Write the number of gametes that form a zygote. _____

Write the number of factors for the pigmentation trait in a zygote. _____

25. On the basis of the foregoing summary of assumptions, infer what must happen to the factors contained in the parent during the production of gametes. _____

_____

---

17. Two; normal and albino.

18. Albinism is masked by normal skin color. (Albinism is recessive.)

19. Yes; because his brother and son are albinos.

20. Two.

21. From his parents.

22. The sperms and eggs unite and the contributions of each parent are contained in the resulting zygote.

23. Two; one.

24. 2; 1; 2; 2.

Definitions

Before going on you need to become familiar with some terminology used in genetics.

P_1 : The parent generation.

F_1 : The first filial generation, or children of the P_1 generation (filial is related to the French *fils* and comes from the Latin *filius, filia*, son, daughter).

F_2 : The second filial generation, or the children of the F_1 generation and the grandchildren of the P_1 generation.

Genotype: The specific factors present in the individual which result in his characteristics. Usually the genotype for any individual characteristic would be represented by two factors as *AA* normal skin color (not albino) or *dd* for a dwarf pea plant. In choosing symbols to represent factors it is customary to use the first letter of the mutant (usually recessive) trait. It is important that the same letter be used for any one trait and that the capital letter be used for the dominant characteristic and the small letter for the recessive.

Phenotype: The appearance of the organism which results from its specific genotype. Thus a genotype of *aa* would result in the phenotype albino.

Pure line (homozygous): All the individuals have only one kind of gene for a trait. If the trait were size, a pure line would contain only factors for tallness (*DD*) or dwarfness (*dd*) but never one gene for each (*Dd*).

Hybrid: A hybrid or heterozygous individual has a pair of contrasting factors for a specific trait, as *Dd*.

25. The pair of factors in the parent for any trait must be halved (from two to one) in its gametes.

■ PROBLEM II

The parents of the following cross are individuals taken from pure lines of snapdragons; one line produced all red individuals and the other produced all white individuals.

P_1 (parents):   Red snapdragon ✕ White snapdragon

Genotype: _____   _____

Gametes: _____   _____

F_1 (first filial generation) phenotype: Pink snapdragon

F_1 genotype: _____

Use the capital letter *R* for one red factor and the small letter *r* for one white factor. Write under the red parent of this cross the number of *R*'s to indicate the number of factors for this trait contained in this individual. These letters represent the genotype of the red parent. Using the small letter *r*, do the same for the white parent. This is the genotype of the white parent.

---

26. In the cross red ✕ white all the F_1 individuals were pink. On the basis of your previous information and assumption, how many factors

for red does the red parent have?_____ How many

Problem II

P_1 genotype:	*RR* ✕	*rr*
Gametes:	*R*	*r*
F_1 genotype:	*Rr*	

factors for white does the white parent have? _____

---

27. Referring back to your previous assumption concerning the history of factors during gamete formation and fertilization, state how many factors must be present in a **gamete** produced by a red parent.

_____ Do the same for the white. _____

26. Two; two.

---

28. When a gamete of a red unites with a gamete of a white snapdragon, what are the factors contained in the zygote? _____

27. One; one.

---

29. Now write under the pink F_1 the letters (symbols) for the factors contained in F_1. This is the genotype of the individual.

If we cross two pink F_1's (pink × pink), a quarter of the offspring are red, half are pink, and a quarter are white, as follows:

28. Red and white.

F_1:	Pink	×	Pink
Genotype:	*Rr*		*Rr*

Gametes: _____ _____   _____ _____

F_2:	¼ red	²⁄₄ pink	¼ white
Genotype:	*RR*	*Rr*	*rr*

---

30. The genotype of the red F_2 individuals is *RR*. Assuming that at least one of these factors (*R*) came from the male parent, where must the other factor (*R*) have come from? _____

29. Gametes: *R r*   *R r*

---

31. The genotype of the white F_2 individuals is *rr*. Where must each of these factors have come from? _____

30. Female.

---

32. Note that the genotype of each pink F_1 (the parents of the F_2 generation) is *Rr*. In the light of the information derived from the answers to the last two questions, state what must happen, during the production of gametes, to the factors that make up the genotype *RR*.

_____

_____

_____

_____

31. One from the male and one from the female.

---

33. Note the ratio of offspring in the F_2 generation. Explain this ratio on the basis of what occurs to the factors during gamete formation and during the union of gametes in fertilization. (In order to do this, you must know that many gametes—often hundreds or even millions—are produced by any one organism and that the union of any

32. They move independently so that they can combine in any of the following combinations:
   R with *R*
   R with *r*
   r with *r*

one sperm and any one egg is determined by chance. Thus any sperm
may unite with any egg. If you make all possible combinations of
gametes that chance will allow, the explanation becomes quite easy.)

_____

_____

_____

_____

■   PROBLEM III

The parents of the following cross are individuals taken from pure
lines of peas. They are called **pure lines** because the tall variety
produces only tall individuals while the short variety produces only
dwarf individuals. (Do not fill in the spaces below until requested to do
so; go on to question **34.**)

P$_1$:   Tall   X   Dwarf

Genotype: _____   _____

Gametes: _____   _____

F$_1$:   Tall

Genotype: _____

_____

34. How many factors for tall does the tall parent have? _____
Use D to represent the tall factor and write in the genotype for the tall
parent (above).

_____

35. How many factors for dwarf does the dwarf parent have?

_____ Use d for the dwarf factor and write in the genotype for
the dwarf parent.

_____

36. How many factors for tall does the F$_1$ have? _____ What

is the basis for your answer? _____

_____

_____

37. Does the F$_1$ have a factor for dwarf? _____ What is the

basis for your answer? _____

_____

_____

33. Each parent has one R and one r.
If they were allowed to combine by
chance, the possible combinations would
be:

          RR 25%
          Rr 50%
          rr 25%

34. Two.

35. Two.

36. One; since the F$_1$ has one factor
for each trait, the tall parent can contribute only one D.

38. Write in the genotype of the F_1. What explanation can you offer to account for the fact that all the F_1's are tall rather than intermediate? _____

_____

If we cross two tall F_1's from the preceding cross, three fourths of the offspring will be tall, and one fourth dwarf, as follows:

38. The factor for tall must mask the factor for dwarf so that it does not appear in the phenotype.

F_1:　　　　　　　　Tall　　　　　X　　　　　Tall

Genotype:　　　　___ ___　　　　　　　___ ___

Gametes:　　　　___ ___　　　　　　___ ___

F_2:　　　　　　　¾ tall　　　　　　　¼ dwarf

Genotypes:　　___ ___　　___ ___　　___ ___　　___ ___

39. Fill in the genotype of the two tall F_1's. When gametes are produced by these tall F_1 parents, how many factors for stature (tall or dwarf) will each gamete contain? _____ Will some gametes contain different factors than others?_____ If so, how many kinds of gametes (containing different factors) will be produced by each parent?

_____

Problem III

Genotype:	*DD*	*dd*
Gametes:	*D*	*d*
F_1 genotype:	*Dd*	

40. Assuming that each parent produces a hundred gametes of the two hundred gametes produced by both parents, how many will there be of each type? _____ What percentage of the total will each type represent?_____

39. One; yes; two kinds.

F_1 genotype:	*Dd*		*Dd*	
Gametes:	*D*	*d*	*D*	*d*
F_2:	*DD*	*Dd*	*Dd*	*dd*

40. 100; 50%.

■ **PROBLEM IV**

As a consequence of the relation of gamete formation and fertilization to the transmission of hereditary factors, many investigations have been made to determine the sequence of events in the process of gamete production. One of the many organisms studied was the worm *Ascaris univalens*. The cells in the gonads of this worm that develop into gametes were found to contain two rod-shaped bodies that stained easily and hence were called **chromosomes** (Greek *chroma*, color). During the formation of gametes these gonadal cells divided. In this process the chromosomes paired, and just before cell division occurred the paired chromosomes separated, one going to one end of the cell and the other going to the other end. When the cytoplasm of the cell divided, it parted between the separated chromosomes. As a consequence, each of the two cells that resulted from the division of the original cell contained only one chromosome. These cells became the gametes. A diagram of this process plus fertilization is given in Figure 25-3.

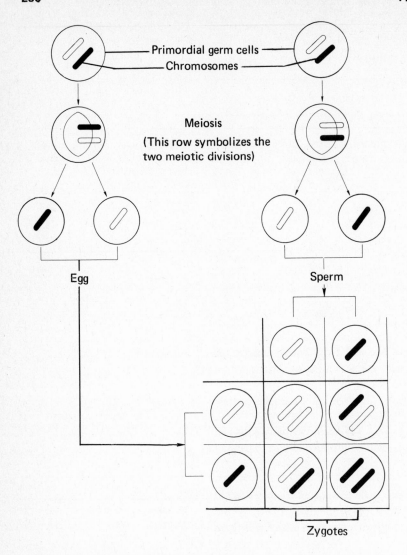

FIGURE 25-3 Reduction of chromosomes in the formation of germ cells, fertilization, and re-establishment of the original chromosome number in the zygote.

41. Compare the history of the imagined hereditary factors in the previous crosses with the observed history of the chromosomes in gamete formation and in fertilization shown in Figure 25-3. State three similarities between the behavior of chromosomes and factors.

a. _____

_____

b. _____

_____

c. _____

_____

42. On the basis of the comparison, make an inference concerning the relation of chromosomes to the supposed hereditary factors. _____

_____

_____

41. a. The number of factors and the number of chromosomes are reduced by one half in gamete formation.

b. The numbers of factors and of chromosomes are brought back to the original number (twice that of the gamete) when the gametes fuse and fertilization takes place.

c. The factors, as well as the chromosomes, are independently assorted and recombine according to chance during fertilization.

Before proceeding to the next question, be sure you have made the comparison of supposed factors and chromosomes and that you realize the significance of this comparison as support for the assumption that hereditary factors exist. The only objects in the cell which can be seen to behave in accordance with the laws of heredity are the chromosomes. The gene is a model. It was invented to help us bridge the gap between the process of heredity and its physical causes.

Try to answer all the rest of the questions during this laboratory period. If you do not finish, complete the rest at home. If you do not understand a question or a procedure, ask your instructor for an appointment to go over your work during office hours. Remember, do not look at the explanation until you have made a real effort to solve the problem yourself.

As you observed in the cross of the tall pea with the dwarf pea, the F_1 generation was all tall. This capacity of one factor to express itself completely, even when the factor for the other characteristic is present, is called **dominance**. The characteristic expressed is known as **dominant**. The characteristic not expressed is called **recessive**.

If both factors for a certain trait are dominant, or both recessive, the individual is said to be **homozygous** for that particular trait. If one factor is dominant and the other is recessive the individual is said to be **heterozygous** for that particular trait.

42. They seem to be identical in behavior.

Drill Questions

43. In tomatoes, red fruit (Y) is dominant over yellow fruit (y). If a homozygous red fruit plant is crossed with a yellow fruit plant:

a. What is the appearance (phenotype) of the F_1 generation?

_____

b. What are the genotypes of the F_2 generation? (Mate two F_1's.)____

_____

c. What is the appearance (phenotypes) of the F_2 generation? _____

_____

44. If hornless (*H*) in cattle is dominant over horned (*h*) and two homozygous cattle, one hornless and the other horned, are crossed, what will be the genotype and the phenotype of the first generation?

_____

_____

_____

43.

P_1:	Red X	Yellow
Genotype:	YY	yy
	(homozygous means that both factors are the same)	
Gametes:	Y Y	y y
F_1 genotype:	Yy	
a. F_1 genotype:	Red (red is dominant over yellow)	
F_1:	Yy X	Yy
Gametes:	Y y	Y y
b. F_2 genotypes:	YY Yy Yy yy	
c. F_2 phenotypes:	¼ red + ²⁄₄ red + ¼ yellow = 3 red, 1 yellow.	

45. If one of these animals were mated to one of the parents and another mated to the other parent, what types of calves could be produced from each cross (phenotypes)? _____

_____

_____

44.

P_1:	HH X	hh
	(each parent is homozygous)	
Gametes:	H H	H h
F_1 genotype:	100% Hh	
F_1 phenotype:	100% hornless (hornless is dominant over horned)	

46. A hornless bull is bred to three cows, A, B, and C. Cow A is horned and produces calf A′, which is horned. Cow B is hornless and produces calf B′, which is horned. Cow C is horned and produces calf C′, which is hornless. Give the genotypes of all seven animals (bull, three cows, three calves). _____

_____

_____

45. Mate the F_1 (*Hh*) with each parent:

	Hh X	HH
	(hornless parent)	
Gametes:	H h	H H
F_2 genotypes:	½ HH and ½ Hh	
F_2 phenotype:	All hornless	
	Hh X	hh
	(horned parent)	
Gametes:	H h	h h
F_2 genotypes:	½ Hh and ½ hh	
F_2 phenotypes:	½ hornless, ½ horned	

47. A brown-eyed man whose mother was blue-eyed marries a brown-eyed woman whose father had blue eyes. What are the chances that this couple will have a blue-eyed child? (Brown is dominant to blue.) _____

_____

_____

46. The hornless bull must have at least one factor for hornlessness, *H*. It may have either *h* or *H* for the other factor. Since we do not know which it is, we will represent the genotype of the hornless bull as *H*?.

Cow A: Cow A is horned and therefore must have *hh*, because if an *H* were present in its genotype it would mask the effect of the *h* and the cow would be hornless, since hornless (*H*) is dominant over horned (*h*).

Cow A   Bull
Genotype of parents: *hh*   *H?*
But calf A′ is horned and therefore must
have *hh*. Therefore we know that one of
the calf's *h* factors comes from the cow,
but the other must have come from the
bull, since the offspring receives one
factor of every pair from each of its
parents. If the bull gave an *h* to its calf,
its genotype must be *Hh*. Check:

	Cow A	×	Bull
	hh		*Hh*
Gametes:	*h h*		*H h*
F₁ :	50% *hh*, 50% *Hh*		

Cow B: Cow B is hornless (*H?*) and
produces calf B′, which is horned (*hh*).
Each of its *h*'s must come from one of
the parents, and since both parents are
hornless, both must have *Hh*. Check:

	Cow B	×	Bull
	Hh		*Hh*
Gametes:	*H h*		*H h*
F₁ :	25% *HH*, 50% *Hh*, 25% *hh*		

Cow C: Cow C is horned (*hh*) and
produces calf C′, which is hornless (*H?*).
Since cow C can only give an *h*, calf C′
must have the genotype *Hh*. Check:

	Cow C	×	Bull
	hh		*Hh*
Gametes:	*h h*		*H h*
F₁ :	50% *Hh*, 50% *hh*		

48. The predigree in Figure 25–4 shows mother (I-1), father (I-2), and five children. Is the characteristic indicated by the shaded square and circle dominant or recessive? _____

_____

How do you know? _____

_____

47. *B* = brown (dominant), *b* = blue; brown-eyed man = *B?*, brown-eyed man's mother = *bb*. Brown-eyed man had to get a *b* from his mother since she did not have a *B*. Therefore, his genotype must have been *Bb*. Brown-eyed woman (*B?*) had a father with blue eyes (*bb*) and therefore must have been *Bb*. Check:

	Brown-eyed	×	Brown-eyed
	man		woman
	Bb		*Bb*
Gametes:	*B b*		*B b*
F₁ :	25% *BB*, 50% *Bb*, 25% *bb*		

(This predicted ratio is not to be considered anywhere near valid unless at least 30–40 offspring are produced. In other words, if the couple had a child with genotype *BB*, this does not mean that you could predict with any degree of certainty that their next three children would be *Bb*, *Bb*, and *bb*. See footnote on page 294.)

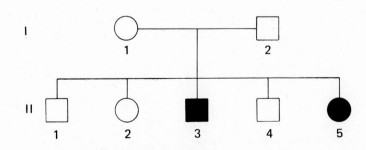

FIGURE 25–4

49. In Figure 25-5, is the characteristic indicated by the shaded

squares and circles dominant or recessive? _____

How do you know? _____

_____

48. The shaded figures indicate recessive factors, since they appear in the offspring (indicating that the parents must have the factors) yet did not show up in the parents. Therefore, the parents must have been $Dd \times Dd$ and the shaded offspring dd (pure recessives). The non-shaded offspring could be either DD or Dd and cannot be determined with certainty.

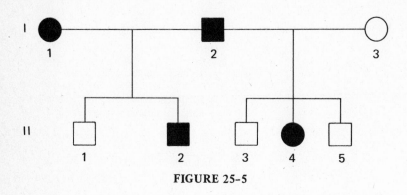

FIGURE 25-5

50. Write the genotypes of the individuals for whom genotypes can be determined below the squares and circles of the pedigree (Figure 25-5).

49. In Figure 25-5 parents I-1 and I-2 produce a child, II-1, which does not exhibit the characteristic. If the shaded squares indicated pure recessives (dd), all the children would have dd (since both parents must have had dd) and would show up as shaded squares. Since this is not the case, shaded squares must indicate a dominant trait.

51. Some types of feeblemindedness apparently are hereditary (see Figure 25-6). Mark each statement (**a** through **p**) with a T if it is true on the basis of facts given, F if it is false on the basis of facts given, or I if there are insufficient data on which to base a true or false answer. (If you fill in the genotype of each individual in the pedigree chart where the genotype can be determined, you will find the questions easier to answer.)

50. The genotypes which can be known with certainty are those containing only pure recessives (dd), which must be the white boxes since the shaded ones possess at least one dominant gene and are either Dd or DD. One of the shaded boxes, I-2, must be Dd since it produces children II-3 and II-5, which are dd. II-4 must be Dd because it shows the dominant trait (shaded) and has a parent which was dd. Answers:

I-1	Dd	II-2	? (DD or Dd)
I-2	Dd	II-3	dd
I-3	dd	II-4	Dd
II-1	dd	II-5	dd

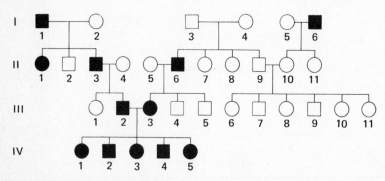

FIGURE 25-6   Pedigree chart of one type of feeblemindedness. The shaded squares and circles represent feeblemindedness, the white squares and circles normal individuals.

a. When both parents are feebleminded, all the children are feebleminded. _____

b. If I-2 had been feebleminded, all of her children would have been feebleminded. _____

c. Feeblemindedness is a dominant characteristic. _____

d. If one of the parents is feebleminded, most of the children will be feebleminded. _____

e. Both I-3 and I-4 were heterozygous. _____

f. The genotype of II-4 can be determined. _____

g. The genotypes of II-7 and II-9 are the same. _____

h. One of the parents of II-4 must have been feebleminded. _____

i. The genotype of II-9 can be determined with a reasonable degree of certainty. _____

j. One of the parents of II-4 could have been heterozygous. _____

k. If II-7 were to marry a normal-appearing man, all her children would be normal. _____

l. II-10 is homozygous. _____

m. III-1 is homozygous. _____

n. All children of II-9 and II-10 are homozygous. _____

o. The genotype of III-2 can be determined. _____

p. If IV-1 should have a normal child, the father of the child would have to be normal. _____

There is a disease of genetic causation which is associated with a blood factor first studied in Rhesus monkeys (hence the factor is called the Rh factor in human blood). This disease, under certain circumstances, causes a frequently fatal anemia in an unborn fetus or a newborn infant. This condition is known as erythroblastosis. When an Rh-negative woman (a woman who does not have the factor) marries an Rh-positive man, the disease may develop in the offspring. However, if the mother is Rh-positive there is no danger that the disease will develop in the fetus. The Rh-negative person is always homozygous (*rr*) for the recessive gene. The Rh-positive person may be either homozygous or heterozygous.

The Rh factor in blood is harmless in itself. It is contained in the blood of 87% of white people and absent from the other 13%. The ratio is only slightly different among the other races.

When an Rh-negative woman carries an Rh-positive fetus, antibodies are produced in the mother's blood against the Rh antigen on the red blood cells of the fetus. (An **antigen** is a proteinaceous substance which stimulates the production of antibodies and reacts with them.) The presence of the antibody in the circulatory system of the fetus produces the disease erythroblastosis, characterized by anemia and jaundice, owing to the destruction of blood cells in the fetus. If the Rh-positive child is a first baby, it may not develop the disease, for the mother has not built up sufficient antibodies, but by the time the second child is born the mother may have built up enough antibodies against the Rh factor to cause the child to develop symptoms. Recent

51.
a. T
b. T
c. F
d. F
e. T
f. T
g. I
h. I
i. T
j. T
k. I
l. F
m. F
n. I
o. T
p. T

improvements in care during birth avoid the presence of fetal red blood cells in the mother's blood, virtually eliminating the disease. A typical pedigree in Figure 25-7 illustrates the transmission of the disease.

I-1, I-2, and II-6 are Rh-positive. All others may or may not be Rh-positive. The shaded squares and circles indicate *the only individuals who could possibly have erythroblastosis,* namely Rh-positive children of Rh-negative mothers. Shaded circles *do not* mean that the individual is Rh-negative; rather he is **the Rh-positive child of an Rh-negative mother.**

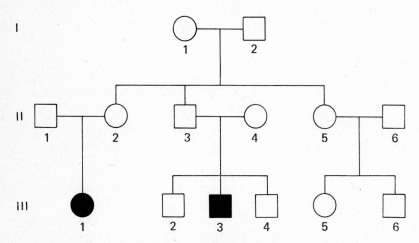

FIGURE 25-7 Pedigree chart of Rh inheritance. Shaded squares and circles mean that the person *could have erythroblastosis.*

52. Determine the genotypes of as many individuals as possible and write them over the circles and squares of Figure 25-7. *Do not look at the solution to this problem until you have spent at least fifteen minutes trying to solve it yourself.*

---

53. Check the correct answer:

a. A disease of genetic causation means that:

(1) The disease is caused by a vitamin deficiency.

(2) The condition which is responsible for the development of the disease is inherited.

(3) There is transmission of an acquired disease.

(4) The disease is caused by bacteria.

b. The blood cells which are involved in the disease of the fetus are:

(1) Red blood cells.

(2) White blood cells.

(3) Platelets.

(4) Germ cells.

c. The child with erythroblastosis is always:

(1) Rh-positive.

(2) Rh-negative.

52. In order to expedite answering this question, fill in the obvious genotypes first. Below will be found the rationale for assigning genotypes. Refer to Figure 25-7 to check your answers.

Given: I-1, I-2, and II-6 are Rh-positive.

Parents I-1 and I-2 must be *Rr* because one of their daughters (II-2) is the mother of a child who can have erythroblastosis. She (II-2) is therefore Rh-negative and has a genotype of *rr*. If one of their children has a genotype of *rr*, parents I-1 and I-2 must each have contributed an *r*. Since it was given that both I-1 and I-2 are Rh-positive, each must have an *R*, making both *Rr*.

II-1 must be *Rr* or *RR* because his child can have erythroblastosis and therefore must be Rh-positive. Since his wife (II-2) is Rh-negative, she is *rr* and cannot contribute an *R* to her child. II-1 must have contributed the *R*.

(3) Homozygous.

(4) A male.

d. Which one of the following must have been correct genotypes of I-1 and 1-2 (Figure 25-7)?

(1) I-1 *Rr* and I-2 *RR*.

(2) I-1 *rr* and I-2 *RR*.

(3) I-1 *Rr* and I-2 *Rr*.

(4) I-1 *RR* and I-2 *RR*.

e. Which one of the following would explain erythroblastosis of III-1?

(1) The mother was *rr*, and the father *rr*.

(2) The mother was *Rr* or *RR*, and the father *rr*.

(3) The mother was *RR*, and the father *Rr* or *RR*.

(4) The mother was *rr* and the father *Rr* or *RR*.

f. The genotype of II-3 may be:

(1) *rr*.

(2) *Rr*.

(3) *RR*.

g. A normal child born after a child with erythroblastosis from the marriage of II-3 and II-4 could be explained by the fact that:

(1) The father was Rh-positive, and the mother Rh-positive.

(2) The father was homozygous for the Rh factor, and the mother Rh-negative.

(3) The father was heterozygous for the Rh factor, and the mother Rh-negative.

(4) The father was heterozygous for the Rh factor, and the mother Rh-positive.

h. In the cousin marriage between III-4 and III-5 (assume III-5 is *RR*) you would expect:

(1) All normal children.

(2) A normal child followed by a child with erythroblastosis.

(3) All children to have erythroblastosis.

(4) Three normal children and one child with erythroblastosis.

---

54. Problem II (page 282) discussed the heredity of red and white flowers which produced pink offspring. Was the gene for either color

dominant over the other? _____

This type of heredity is known as **incomplete dominance**. It is illustrated in question **55**.

II-2 is *rr*, as explained above.

II-3 is *Rr* because he is the father of an Rh-positive child (III-3) and therefore must have contributed an *R*. His other two children (III-2 and III-4) cannot have erythroblastosis even though their mother is Rh-negative; therefore, they must be Rh-negative, too. This would make them *rr*.

If one of his children is *Rr* and the other two are *rr*, parent II-3 must be *Rr*.

II-4 must be *rr* since she is the mother of a child who can have erythroblastosis.

III-1 is susceptible to erythroblastosis and therefore is the Rh-positive child of an Rh-negative mother. Her genotype must be *Rr*, since she can get only *r* from her mother.

III-2 and III-4 are both *rr* since they are the children of an Rh-negative mother and cannot get erythroblastosis.

III-3 is *Rr* since he is susceptible to erythroblastosis. His mother, being a pure recessive, can give him an *r* only.

In question 53h, you are told to assume that III-5 is *RR*.

53.

a. 2

b. 1

c. 1

d. 3

e. 4

f. 2

g. 3

h. 1

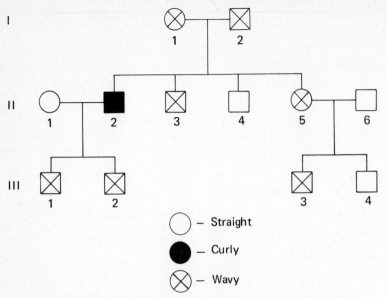

FIGURE 25-8 Pedigree chart of the inheritance of hair curliness.

55. The pedigree in Figure 25-8 shows the probable inheritance of hair texture. Mark each statement (**a** through **j**) with a 1 if the statement is true, 2 if the statement is false, 3 if there is a 25% chance that the statement is true,* 4 if there is a 50% chance that the statement is true,* or 5 if no decision is possible because of insufficient data for a conclusion.

54. No.

a. Wavy hair is dominant.   _____

b. Individual I-1 is homozygous.   _____

c. If individual II-4 married a woman with straight hair all of the offspring would have straight hair.   _____

d. Individual II-2 is homozygous.   _____

e. If II-2 married someone with wavy hair, the first child would have wavy hair.   _____

f. One of the parents of I-2 had the same genotype as I-2.   _____

g. If II-1 and II-2 had had another child it would have had curly hair.   _____

h. The genotype of II-3 can be determined on the basis of this pedigree only and not by looking at the person.   _____

i. The genotype of II-4 can be determined on the basis of this pedigree only and not by looking at the person.   _____

j. Curly hair is recessive.   _____

*It should be understood that one can predict the chances of the distribution of genetic factors only in large populations where the laws of chance can operate. For example, if you flip a coin 100 times, you can predict that close to 50 of the tosses will be heads and nearly 50 will be tails. However, when you predict that the first toss will be a head, there is much less likelihood that your prediction will be successful: you might very well toss a coin five times in a row and get all tails and no heads. Predictions of this sort are more valid as you increase the number of trials. There is little likelihood of accuracy in predicting that one out of four children will have blue eyes or curly hair. If we were referring to fish, however, which can have hundreds or thousands of offspring, we could predict the distribution of characteristics in their offspring with more validity. All predictions you are asked to make concerning the ratio of human and other offspring where the parents produce few young are only intended to show that you understand the mechanisms of genetics. They are not to be considered valid.

In most cases of heredity the heterozygous phenotype is intermediate between the two parents. Incomplete dominance then, is the rule rather than the exception. For example, a human genetic disease involving incomplete dominance which has been studied in detail in recent years is sickle cell anemia. The gene is found in the black population. It produces a modified hemoglobin molecule, which causes red blood cells to be fragile and to clump together. The cells also appear curved like sickles, rather than as flat discs. In the homozygous condition, normal emotional or physical stress causing constriction of the capillaries is often enough to provoke a painful episode where the red blood cells clump together at the junctions of the capillaries.

The heterozygote has sickle-shaped cells, but only mild symptoms.

---

56. From the above evidence, is the gene for sickle cell anemia

completely dominant? _____ Evidence which led to this conclusion is the fact that the trait shows up in the heterozygote, but the presence of the normal gene causes a modification of the sickling

trait, demonstrating that this is an example of _____

55. Genotypes are straight hair *SS*, wavy hair *Ss*, curly hair *ss*.

a. 2	d. 1	g. 3
b. 2	e. 4	h. 2
c. 1	f. 5	i. 2
		j. 2

A man decided to raise rabbits. He found a variety which he liked, a mottled black and white. He bought a male and female and bred them. The litter consisted of three mottled rabbits and one white one. On the basis of these data he made a hypothesis regarding the manner of inheritance of the characteristic he desired (mottled).

57. What hypothesis would you make on the basis of these data?

☐ Hypothesis: _____

56. No (it is dominant, otherwise the trait would not have shown up in the heterozygote, but not completely dominant, otherwise the heterozygote would not have shown a mild form of the disease); incomplete dominance.

---

58. Were the rabbits he bought heterozygous or homozygous? _____

57. Mottled is dominant over white.

---

59. How do you know? _____

58. Heterozygous.

---

60. Would you expect all mottled offspring to be heterozygous?

Explain. _____

59. If they were homozygous, all offspring would have been mottled.

---

61. Since he wanted to raise a pure (homozygous) race of mottled rabbits, he bought, on the basis of his first cross, twenty more mottled females. Would you have done this on the basis of your conclusions?

_____ Why or why not? _____

60. No, if both parents were heterozygous and chance was operating, they could produce one pure homozygous dominant, two hybrids and one homozygous recessive.

62. He bred the twenty females to the male and obtained 54 black, 97 mottled, and 49 white. Was your tentative conclusion valid? _____

Why or why not? _____

_____

_____

61. No, because he had not determined which of the mottled rabbits was hybrid and which was pure dominant.

63. State another hypothesis which will cover the data completely.

☐   Hypothesis: _____

_____

62. No; now that black is introduced, we see that our sample was too small to reveal another factor, and that we are dealing with incomplete dominance.

64. How do you explain the discrepancy between the data from the first cross on which you based your first hypothesis and the data from the offspring of the twenty rabbits? _____

_____

_____

_____

63. Question **62** states that the mottled male was bred to 20 mottled females, resulting in a ratio of offspring of approximately 1 black: 2 mottled: 1 white. This indicates that both parents were heterozygotes and that coat color in this case is an example of incomplete dominance. (BB = black, Bb = mottled, bb = white)

P_1:	Bb	X	Bb
Gametes:	B b		B b
F_1:	BB Bb		Bb bb
Phenotypes:	1 black, 2 mottled, 1 white		

65. Is your second hypothesis likely to be valid? _____

64. Our sample was too small for chance to operate, and the BB combination did not appear.

66. Why or why not? _____

_____

_____

65. Yes.

66. The sample is large enough now, containing 200 individuals.

The Back-cross

A **mutation** (factor change) appeared in one member of a flock of sheep. He had very short, stocky legs, and the farmer who owned him had no fear that this sheep might jump fences and leave the pasture. The farmer realized that a flock of short-legged sheep would be an advantage. He wished to determine whether or not the characteristic was hereditary, so he bred this male to a normal female. All of the offspring had short legs.

67. Was the characteristic dominant or recessive? _____

Give the genotype of the lambs. _____

68. He decided to raise a pure line (homozygous line) of these sheep. What would be the genotype of individuals homozygous for this

67. Dominant; Ss.

characteristic? _____

69. He bred the short-legged ram to several ewes and then mated short-legged males to short-legged females. Give the crosses between a homozygous short-legged ram mated to (1) a heterozygous short-legged ewe and (2) a homozygous short-legged ewe.

68. *SS.*

70. How would the farmer determine which of these short-legged

sheep were homozygous? _____

_____

69. S = short-legged, s = normal. The ram has a genotype of *SS* because when mated to several normal ewes (ss) he produced only short-legged (Ss) offspring. When the short-legged rams were mated with the short-legged ewes, the possible offspring are determined as follows:

	Ram	Ewe 1	Ewe 2
	SS	*Ss*	*SS*
Gametes:	*S S*	*S s*	*S S*
F_1:		50% *SS*	100%
		50% *Ss*	*SS*
Phenotypes:		100% short-legged	100% short-legged

71. How would he develop his pure line of short-legged individuals? (Remember that his pure line must never produce any of the long-legged type.) _____

_____

70. The farmer could test to see whether his sheep were *SS* or *Ss* by a back-cross. This is done by mating the animal which is suspected of being heterozygous with a pure recessive. If the animal being tested is heterozygous, the recessive gene will show up in some of its offspring which will be pure recessive. For example:

	Ss (heterozygous) ×	*ss*
Gametes:	*S s*	*s s*
F_1:	50% *Ss*, 50% *ss*	
	50% of offspring will be normal.	

But

	SS (homozygous) ×	*ss*
Gametes:	*S S*	*s s*
F_1:	100% *Ss*	

If no normal lambs appear, the animal being tested must be homozygous dominant.

SECTION B. TWO-FACTOR CROSSES

In Section A, it was learned that by hypothesizing the presence of factors, hereafter called **genes**, and the relationship that exists between them and certain characteristics, many facts of biology could be logically related. At this point many other questions arise. Is all inheritance gene-controlled, or are there other mechanisms? How do genes for different characteristics affect one another in transmission and in activity? Can all inheritance be reduced to a simple pattern? This section answers some of these questions, at least in part, and provides a basis for understanding more facts about inheritance.

■ PROBLEM V

In some crosses it is possible to trace the history of two pairs of characteristics. In the following pea cross, the tall parent that produced smooth seed coats was obtained from a strain that had always produced tall offspring with smooth seed coats. The other parent came from a strain that had always produced dwarf offspring with wrinkled seed coats. When these two were crossed, the F_1 was tall and smooth. The F_2 produced by mating two F_1's showed four combinations—tall and smooth, dwarf and wrinkled, tall and wrinkled, dwarf and smooth.

72. Determine the genotype of each individual of P_1 and F_1 and F_2 and write it in the proper place in the cross.

P_1:   Tall, smooth   X   Dwarf, wrinkled

Genotype: _____   _____

Gametes: _____   _____

F_1 Tall, smooth

Genotype: _____

Two F_1's
crossed: Tall, smooth   X   Tall, smooth

Gametes: _____   _____

F_2:   Tall,   Tall,   Dwarf,   Dwarf,
                smooth   wrinkled   smooth   wrinkled

Genotypes: _____

            _____   _____   _____

            _____   _____   _____

            _____

73. Which of the characteristics are dominant and which recessive?

_____

_____

Note, in the cross just completed, that one P_1 individual was tall and smooth; that is, the characteristics tall and smooth were in the same individual. Note also that the other P_1 was dwarf and wrinkled; that is, the characteristics dwarf and wrinkled were in the same individual. The F_1 individuals were tall and smooth, and the dwarf and wrinkled factors were not expressed. In the F_2, tall and smooth are again combined in some individuals and dwarf and wrinkled are combined in others, but there are also two new types of individuals–tall, wrinkled and dwarf, and smooth. These individuals show combinations of characteristics not found in the P_1. The problem is to explain the appearance of these new combinations. Remember that for each trait (for example, height and seed-coat condition) there are two genes; hence the P_1 tall, smooth individual had a total of four genes, two for each trait. The same is true of the P_1 dwarf, wrinkled individual. Likewise the F_1 tall, smooth individual had four genes.

72.	$DDWW$	X	$ddww$
Gametes:	DW		dw
F_1:		$DdWw$	
F_1's crossed:	$DdWw$	X	$DdWw$
Gametes:	Dw Dw		Dw Dw
	dW dw		dW dw

F_2:

Tall, smooth Tall, wrinkled
1 $DDWW$ 1 $DDww$
2 $DDWw$ 2 $Ddww$
2 $DdWW$
4 $DdWw$

Dwarf, smooth Dwarf, wrinkled
1 $ddWW$ 1 $ddww$
2 $ddWw$

---

74. By imagining the possible fates or paths of each of these genes in the cross of two F_1 individuals, explain the appearance of the four

different combinations of characteristics in the F_2. _____

_____

_____

73. Tall and smooth are dominant, and dwarf and wrinkled are recessive.

In the previous cross of tall pea with dwarf pea, in which only one pair of characteristics was involved, we hypothesized a relation between genes and chromosomes. (See questions **34–40**.) Does the same relationship occur in a cross involving two pairs of characteristics? To answer this question, investigations were made on many organisms having more than one pair of chromosomes. A modified chromosome history of the worm *Ascaris lumbricoides* is given in Figure 25-9 (page 300). Study this and compare it with the history of the genes for tall, dwarf, wrinkled, and smooth in peas.

75. List four ways in which the behavior of chromosomes during gamete production and fertilization is similar to the behavior of the genes in the crosses. (You have already listed three similarities in question **41**, page 286.)

74. There are nine cases where at least one dominant gene for each trait is present in the F_2 genotypes. These range from the most common, $DdWw$, to the least common, the pure dominant $DDWW$. There are three cases each where the dominant gene for one trait is present, but the other trait is represented by recessive genes only, as $Ddww$ or $ddWw$. Finally, there is only one combination of pure recessive genes for both traits, $ddww$. This gives us the classical dihybrid cross ratio of phenotypes, 9:3:3:1.

a. _____

_____

b. _____

_____

c. _____

_____

d. _____

_____

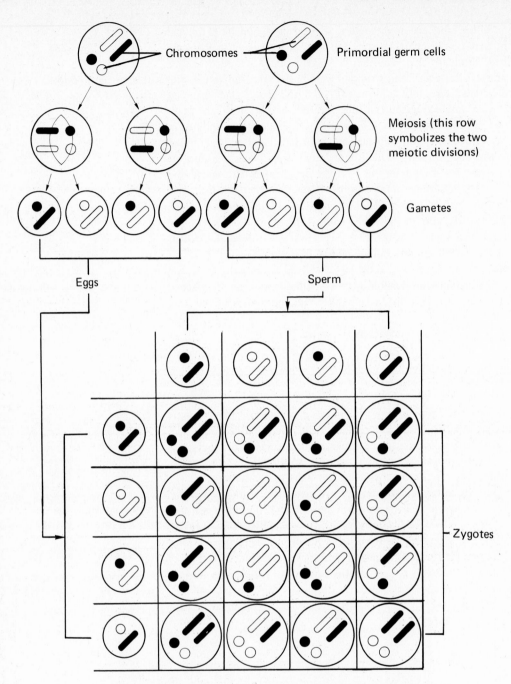

FIGURE 25-9 Diagram of gamete formation and fertilization.

76. What additional evidence is derived from the comparison of the behavior of two pairs of genes and the behavior of two pairs of chromosomes in gamete formation and fertilization to support your original hypothesis in question **42** (page 287), concerning the relationship between chromosomes and genes? _____

_____

_____

75. a. The number of genes and the number of chromosomes are reduced by one half in gamete formation.

b. The numbers of genes and of chromosomes are brought back to the original number when the gametes fuse and fertilization takes place.

c. The genes, as well as the chromosomes, are independently assorted and recombine according to chance during fertilization.

d. Only one gene and one chromosome of every pair go to each offspring.

77. What will be the proportion of tall and wrinkled, tall and smooth, dwarf and smooth, and dwarf and wrinkled offspring in the F_2 generation of the pea cross presented in question 72? In order to determine this consider that two parents in Figure 25–9 (primordial germ cells) represent the two F_1 individuals. Place on one chromosome of each parent the symbols W and D. On the other chromosome of each parent place w and d. This corresponds to a genotype of $WwDd$. Now continue to label the chromosomes throughout the diagram. You will have the possible genotype combinations of the F_2 in the large squares designated "zygotes."

76. During meiosis the chromosomes divide and one member of each pair goes to each gamete. Similarly, one gene of each pair goes to each gamete. In two factor crosses, the chromosomes and the genes can be shown to move together; they assort independently, but the genes are always associated with the same chromosomes. This substantiates the idea that genes are on (or part of) chromosomes.

---

78. What will be the appearance of each zygote after it develops? Write in each square what the appearance of that individual will be.

77.

1	*DDWW*	1	*DDww*
2	*DDWw*	2	*Ddww*
2	*DdWW*		
4	*DdWw*	1	*ddWW*
		2	*ddWw*
		1	*ddww*

---

79. In what proportion will each phenotype appear in the F_2?

78. 9 tall, smooth
3 tall, wrinkled
3 dwarf, smooth
1 dwarf, wrinkled

_____

_____

_____

---

79. See answer to question 78.

■ **PROBLEM VI**

This problem deals with crosses involving two pairs of genes in each parent. In man, brown eyes (B) are dominant over blue eyes (b), and right-handedness (L) is usually assumed to be dominant over left-handedness (l). If a homozygous brown-eyed, right-handed man marries a blue-eyed, left-handed woman, what will be the appearance of the F_1? What will be the genotype? (Key: B = brown, b = blue, L = right-handedness, l = left-handedness.)

Parents:	*BBLL*	X	*bbll*
Gametes:	*BL*		*bl*
F_1 genotype:		*BbLl*	
F_1 phenotype:		Brown-eyed, right-handed	

If this individual, the F_1, marries a person of identical genotype with regard to eye color and handedness, what will be the possible offspring in the F_2? In other words, if an individual who is heterozygous for two traits marries an individual with the same genotype, what genotypes might arise from that cross?

F_1:	*BbLl*	X	*BbLl*
Gametes:	*BL*		*BL*
	Bl		*Bl*
	bL		*bL*
	bl		*bl*

The following checkerboard is used to obtain the possible genotypes of the individuals in the F_2. The gametes are placed at the top and side and the combinations in the various squares. For example, the upper left square represents the result of a sperm containing the *B* and *L* factors meeting with an egg containing the *B* and *L* factors.

Sperm Factors

Egg Factors	*BL*	*Bl*	*bL*	*bl*
BL	*BBLL*	*BBLl*	*BbLL*	*BbLl*
Bl	*BBLl*	*BBll*	*BbLl*	*Bbll*
bL	*BbLL*	*BbLl*	*bbLL*	*bbLl*
bl	*BbLl*	*Bbll*	*bbLl*	*bbll*

F_2 phenotypes: 9 brown-eyed, right-handed
3 brown-eyed, left-handed
3 blue-eyed, right-handed
1 blue-eyed, left-handed

Drill Questions

80. A blue-eyed, left-handed woman marries a brown-eyed, right-handed man who is heterozygous for both traits. What are the possible genotypes and the phenotypes of the offspring? ________________

____

81. A brown-eyed, right-handed man marries a blue-eyed, right-handed woman. Their first child is blue-eyed and left-handed. What are the genotypes of the parents? ________________

____

80. *B* = brown, *b* = blue, *L* = right-handed, *l* = left-handed.
Given: Man is heterozygous for both traits.

	Woman	×	Man
	bbll		*BbLl*
Gametes:	*bl*		*BL Bl bL bl*

	BL	*Bl*	*bL*	*bl*
bl	*BbLl*	*Bbll*	*bbLl*	*bbll*

Genotypes:
25% *BbLl*, 25% *Bbll*, 25% *bbLl*, 25% *bbll*

Phenotypes:
25% brown-eyed, right-handed
25% brown-eyed, left-handed
25% blue-eyed, right-handed
25% blue-eyed, left-handed

81. Since the child is a pure recessive, *bbll*, the father must be *BbLl*; otherwise he would not have any recessives to give to his child. The mother must be *bbLl*.

In the coat color of shorthorn cattle there is lack of dominance. The gene for red coat is designated (*W*) and the gene for white (*w*). The heterozygous condition gives a roan (brown and white) coat. Hornless (*H*) is dominant over horned (*h*).

Do questions 82–84 before checking the answers, which follow question 84.

82. If a homozygous horned white bull is bred to a homozygous

hornless red cow, what will be the phenotype of the F_1? _____

_____

Of the F_2? _____

_____

83. If one of the F_1 bulls is bred to the parental cow, what will be

the phenotypes of the offspring? _____

84. If one of the F_1 cows is bred to the parental bull, what will be

the phenotypes of the offspring? _____

_____

In tomatoes red fruit is dominant over yellow and tall vines are dominant over short vines. A tomato breeder has pure (homozygous) races of yellow-fruited tall plants and red-fruited short plants. He wants a pure race of red-fruited tall plants.

Answers **85–89** follow question **89**.

85. If he crosses the two races he has, what will be the appearance

of the F_1? _____

_____

86. Will the F_1, when self-crossed, produce the desired pure race? ____

_____

87. What proportion of the F_2 will have the desired phenotype? ____

_____

88. What proportion of the F_2 will be homozygous? _____

_____

89. How can the homozygous plants be determined? _____

_____

_____

82. $hhww$ ✕ $HHWW$
Gametes: hw HW
F_1: 100% $HhWw$
Phenotype: 100% hornless roan cattle

Mate two F_1's $HhWw$ ✕ $HhWw$
to get F_2
Gametes: HW Hw HW Hw
  hW hw hW hw

	HW	Hw	hW	hw
HW	$HHWW$	$HHWw$	$HhWW$	$HhWw$
Hw	$HHWw$	$HHww$	$HhWw$	$Hhww$
hW	$HhWW$	$HhWw$	$hhWW$	$hhWw$
hw	$HhWw$	$Hhww$	$hhWw$	$hhww$

Phenotypes:
  3 Hornless red   1 horned red
  6 hornless roan   2 horned roan
  3 hornless white   1 horned white

Note that the ratio of hornless to horned is 3:1 because hornless is dominant to horned. The ratio of red to roan to white is 1:2:1 because of incomplete dominance.

83. $hHWw$ ✕ $HHWW$
Gametes: HW Hw hW hw HW

F_1: $HHWW$ $HHWw$ $HhWW$ $HhWw$
Phenotypes: 50% hornless red, 50% hornless roan

84. $HhWw$ ✕ $hhww$
Gametes: HW Hw hW hw hw
F_1: $HhWw$ $Hhww$ $hhWw$ $hhww$
Phenotypes: 25% hornless roan, 25% hornless white, 25% horned roan, 25% horned white.

90. A right-handed, blue-eyed man whose mother was left-handed marries a right-handed, brown-eyed woman whose mother was left-handed and blue-eyed. What are the possible appearances of their

children? _____

_____

In humans, a certain type of deaf-mutism is recessive. Let (D) be normal and (d) be deaf-mute. A normal right-handed man married a normal right-handed woman. They had a child who was a left-handed deaf-mute.

Answers **91–93** follow question **93**.

91. What were the genotypes of the parents? _____

92. What was the genotype of the child? _____

93. What were the chances that their only child would be a

left-handed deaf-mute? _____

94. A freckled blue-eyed man married a brown-eyed woman without freckles. They had six children, two blue-eyed without freckles, one blue-eyed with freckles, two brown-eyed with freckles, and one brown-eyed without freckles. Give the genotypes of the parents and of the children. (Freckles are dominant to no freckles.)

_____

_____

_____

_____

85–89. Y = red, y = yellow, S = tall, s = short.

	$yySS$	X	$YYss$
Gametes:	yS		Ys
F_1:	$YySs$		

Phenotype: 100% red tall plants

In order to determine the homozygous red-fruited tall plants, he should mate all red-fruited tall plants to yellow-fruited short plants (pure recessives). Any of the offspring of this mating showing recessive traits must have received a recessive gene from the red, tall plant, making it a hybrid.

90. L = right-handed, l = left-handed, B = brown-eyed, b = blue-eyed.

The man is right-handed, blue-eyed ($L?bb$). His mother was left-handed (ll). Therefore, the man had to get an l from his mother and is $Llbb$.

His wife is right-handed and brown-eyed ($R?B?$). Her mother is left-handed and blue-eyed ($llbb$). Since her mother had only recessive genes, the wife has a genotype of $LlBb$.

	Man	X	Wife
	$Llbb$		$LlBb$
Gametes:	Lb lb		LB Lb lB lb

	Lb	lb
LB	$LLBb$	$LlBb$
Lb	$LLbb$	$Llbb$
lB	$LlBb$	$llBb$
lb	$Llbb$	$llbb$

Phenotypes:
  3 right-handed, brown-eyed
  3 right-handed, blue-eyed
  1 left-handed, brown-eyed
  1 left-handed, blue-eyed

91–93. Both parents must have been heterozygous for both traits ($LlDd$) since their child had to have a genotype of $lldd$ because both recessive traits appeared. The chances of the pure recessive showing up in a dihybrid cross are 1 in 16.

SECTION C. SEX LINKAGE

In April 1910 Thomas Hunt Morgan, the father of fruit-fly genetics, found a white-eyed male fly in a colony of thousands of red-eyed flies. He carefully removed the fly from the colony and mated it with a normal, red-eyed female in a separate container. The bottle was soon populated with the F_1 descendants of this mating, all of which had red eyes.

95. Explain how a white-eye $\times$ red-eye cross could result in 100%

red eyes in the F_1. _____

_____

_____

_____

Next, Morgan allowed the F_1 flies to interbreed. The resulting F_2 generation contained 3470 red-eyed and 782 white-eyed flies.

96. Is this the result you would have predicted? _____

What is the expected ratio? _____

Upon examining the offspring, Morgan began to find clues as to why the F_2 population was so far from the predicted ratio. At first he thought that although the sample was quite large, there was simply a peculiar chance distribution of characteristics. Then he noted that all 782 white-eyed flies were males. The most logical hypothesis was that the flies were somehow incapable of producing white-eyed females. To test this hypothesis Morgan mated the original P_1 white-eyed male to some of his daughter F_1 red-eyed females. The results were 129 red-eyed females, 132 red-eyed males, 86 white-eyed males, and 88 white-eyed females.

97. Was Morgan's hypothesis correct? _____

98. What was the ratio of white-eyed males to white-eyed females in

this cross? _____

99. Could females be white-eyed? _____

---

94. Parents' genotypes: Father $F?bb$ $\times$ Mother $ffB?$.

Since the parents had children with blue eyes (bb) and children without freckles (ff), this means that each parent had recessive genes to contribute to the offspring. The father's genotype was therefore $Ffbb$ and the mother's $ffBb$.

	$Ffbb$	$\times$	$ffBb$
Gametes:	Fb fb		fB fb

	Fb	fb
fB	$FfBb$	$ffBb$
fb	$Ffbb$	$ffbb$

The genotypes of the children are:
2 $bbff$ blue-eyed, without freckles
1 $bbFf$ blue-eyed, with freckles
2 $BbFf$ brown-eyed, with freckles
1 $Bbff$ brown-eyed, without freckles

The purpose of combining the gametes in this case is not to determine the ratios of the offspring, but rather to find out what possible kinds of offspring could result from the mating of the parents. Since a small sample of six offspring does not permit chance to operate, the distribution of genotypes of the six offspring could not be determined precisely, and were instead given in the statement of the problem.

---

95. This would be the expected result of a cross between a pure recessive white-eyed fly and a homozygous red-eyed fly. All of the F_1's would be hybrids, $WW \times ww \rightarrow 100\%$ Ww.

---

96. No; 3:1 or 3000:1000. With a sample this large, the results should have been closer to the expected (chance) ratio.

---

97. No.

---

98. 1:1

100. Did the original F_2 generation have any white-eyed females?____   99. Yes

---

101. Was there any difference in genotypes of the F_1 cross which   100. No.
produced no white-eyed females and the cross between the P_1 male and

the F_2 females? _____ Explain. _____

_____

_____

_____

From these data, Morgan was able to determine that the inheritance   **101.** Yes; the F_1 × F_1 cross mated
of white eye color was somehow associated with sex determination.   two red-eyed hybrids, while the P_1 × F_1
                                                                           cross mated a white-eyed pure recessive
102. What ultimately determines eye color inheritance? _____   male with a red-eyed hybrid female,
                                                                           resulting in the following:
                                                                           $ww \times Ww \rightarrow Ww\ \ Ww\ \ ww\ \ ww$

---

103. What ultimately determines the inheritance of sex?_____   102. Genes.

---

104. Where are these determining factors located in the nucleus of   103. Genes.

the cell?_____

---

105. How can you explain why sex and eye color appeared to go   104. On the chromosomes.
together, yet many other traits appear to be randomly assorted and

yield the expected Mendelian ratios? _____

_____

_____

_____

The association of genes on the same chromosome is called linkage,   **105.** The genes for eye color and the
and when the genes are found on the chromosomes which determine   genes for sex determination must have
sex, this is referred to as **sex linkage**.   been on the same chromosome.

The chromosomes containing the genes which determine whether a
human will be a male or female are the only pair which are not alike in
appearance of the 23 pairs of chromosomes found in human cells. The
chromosome containing the factors which result in femaleness is called
the X chromosome. A genotype of XX will result in a female. The
chromosome with genes for maleness is the Y chromosome. It is smaller
than the X chromosome and contains fewer genes. The genotype of
male is XY. Figure 25–10 represents the male and female chromosomes.

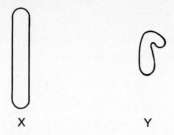

FIGURE 25-10 Schematic drawing of the male (Y) and female (X) chromosomes.

106. In Figure 25-11, the gene for eye color is represented on the X chromosome. Is white eye dominant or recessive? _____

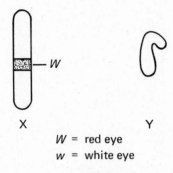

W = red eye
w = white eye

FIGURE 25-11

The Y chromosome has fewer genes on it than the X chromosome. One of the genes missing from the Y chromosome is the gene for eye color in *Drosophila*.

For sex-linked genes the convention is to represent the sex chromosomes and the genes linked to them together, for example, X^W. The male (Y) chromosome lacks many of the genes found on the X chromosome. In this case the Y chromosome would be represented without a letter for a gene for eye color, thus: Y.

107. Write the genotype of a female homozygous for red eye color.

_____

108. Write the genotype of a female heterozygous for red eye color.

_____

109. Write the genotype of a male with red eyes. _____

106. Recessive.

107. $X^W X^W$

108. $X^W X^w$

110. Write the genotype of a white-eyed male. _____   **109.** $X^W Y$

---

111. We are now in a position to calculate the ratios we can expect if populations are large enough so that chance can operate. What would you expect to be the ratio of males to females born in any population?

110. $X^w Y$

---

112. Use the checkerboard below to test your hypothesis:

111. 1:1

Was your hypothesis correct? _____

---

113. Now mate a homozygous red-eyed female fly to a white-eyed male. Remember, the Y chromosome does not have a gene for eye color. Use the symbol X^W for red eye color and X^w for white eye.

112.

	X	Y
X	XX	XY
X	XX	XY

	Homozygous red-eyed female	X	White-eyed male

Genotype: _____   _____

Gametes: _____   _____

                _____   _____

Genotypes: _____   _____   _____   _____

Phenotypes: _____

---

114. Were there any white-eyed males produced? _____

113.

P$_1$ genotypes: $X^W X^W$ $\times$ $X^w Y$

Gametes: X^W X^w X^W Y

F$_1$ genotypes: $X^W X^w$ $X^W X^w$
$X^W Y$ $X^W Y$

Phenotypes: 50% red-eyed females
50% red-eyed males

---

115. Why or why not? _____

114. No.

116. Explain why the cross that Morgan studied, between a white-eyed male and a red-eyed female, could produce white-eyed females. Use the space below for your calculations and list your ratios of genotypes and phenotypes in the box.

Genotypes
Phenotypes

Sex linkage occurs in humans. Genes for both red-green color-blindness and hemophilia (a disease in which the blood fails to clot or clots very slowly) are found on the X chromosome, while the Y chromosome does not have any genes for these traits.

116.

P_1 genotype: $\quad X^w \; Y \quad \times \quad X^W X^w$

Gametes: $\qquad X^w \; Y \qquad\quad X^W \; X^w$

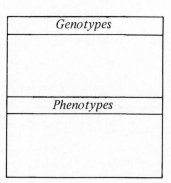

	X^w	Y
X^W	$X^W X^w$	$X^W Y$
X^w	$X^w X^w$	$X^w Y$

Genotypes: $\quad 25\% \; X^W X^w$, $25\% \; X^w X^w$,
$\qquad\qquad 25\% \; X^W Y$, $25\% \; X^w Y$

Phenotypes:
   25% red-eyed females
   25% white-eyed females
   25% red-eyed males
   25% white-eyed males

117. Solve the following problem: A woman with normal vision whose father was colorblind married a colorblind man. What would be the expected ratio (if chance were to operate) of the traits for sex and colorblindness in their children? Place the genotypes and phenotypes in the box.

Genotypes
Phenotypes

---

117. The woman had normal vision so she must have had one X^C. Her father could have given her only his X chromosome (if he gave her his Y chromosome, she would have been a boy). But he was colorblind, so he gave her X^c. Her genotype, therefore, was $X^C X^c$. She married a colorblind man ($X^c Y$).

P_1 genotypes: $X^C X^c$ × $X^c Y$

Gametes: X^C X^c X^c Y

	X^c	Y
X^C	$X^C X^c$	$X^C Y$
X^c	$X^c X^c$	$X^c Y$

Genotypes: 25% $X^C X^c$, 25% $X^c X^c$, 25% $X^C Y$, 25% $X^c Y$

Phenotypes:
  25% normal-visioned girls
  25% colorblind girls
  25% normal-visioned boys
  25% colorblind boys

26 Analysis of a Famous Investigation

Discovery of the Structure of the DNA Molecule

In 1968, a book was published of singular importance to the history of science. It was a personal account of the discovery of the structure of the DNA molecule—perhaps the most important biological discovery of this century.

The book is entitled *The Double Helix;* it was written by James D. Watson,* one of the three persons who shared the Nobel Prize for the discovery of the structure of DNA. When it was published, the book almost immediately became a bestseller, testifying to the lucid and personal style of its author, who was able to convey the excitement of the discovery and the personal interactions that make this a human adventure rather than a dry scientific report. You will do well to read the book in its entirety, for it provides an unusual opportunity for students of biology; rarely is the development of an important idea clearly laid out for us to follow.

The essence of the book comprises this exercise. While attempting to put together your own DNA molecule you will have the opportunity to follow some of the thoughts and ideas which characterized the search for this all-important molecule—a molecule so critically important that to understand the structure of DNA is to comprehend the transmission of life from parent to offspring. It has been said that the discovery of the way in which DNA functions is more potentially dangerous to mankind than the discovery of the methods of releasing energy from the nucleus of the atom. It has also been said that the promise of understanding the fundamental way in which life replicates itself will help us to cure cancer, perhaps even to do away with genetic defects, creating better humans. For better or for worse, the gene has been discovered; this is the story of that discovery.

PRELIMINARY INFORMATION

In 1944 Erwin Schrödinger, a theoretical physicist, wrote a book called *What Is Life?* In it he described genes as special types of protein molecules. Almost at the same time a bacteriologist, O. T. Avery, was carrying out experiments which showed that hereditary traits could be transmitted from one bacterial cell to another by purified DNA molecules. This suggested that Schrödinger was wrong and that genes were composed of DNA, not proteins.

Maurice Wilkins, an X-ray crystallographer and physicist in England, was one of the few persons able to see the importance of comprehending the structure of DNA. In his laboratory was a brilliant and irritating young X-ray crystallographer named Rosalind Franklin. Together (or, rather, simultaneously) these two scientists began to try to photograph molecules of DNA

*James D. Watson, *The Double Helix,* Atheneum, New York, 1968, 226 pp.

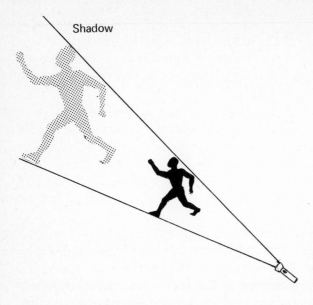

Shadow

FIGURE 26-1 Casting a shadow with a light source is similar to X-ray crystallographic photographs in that the "shadow" is much larger than the original.

by passing a stream of X-rays through the molecule and catching the "shadow" of its components on photographic paper, creating a "picture" of the molecule. See Figures 26–1 and 26–2.

Francis Crick, another physicist, became aware of the importance of DNA, but he was caught up in the study of proteins and did not have the stimulation of someone else interested in DNA to cause him to switch his interests as Wilkins did.

Another important actor in the drama was the American Linus Pauling ("the greatest of all chemists"), who had already won a Nobel Prize for his work. Pauling began the study of the structure of DNA in his laboratory in California.

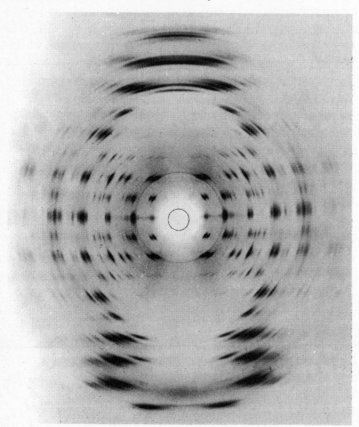

FIGURE 26-2 An early X-ray photograph of the crystalline structure of DNA. Note the black spots, which are shadows of the molecules making up the larger DNA molecule. The individual molecules which are too tiny to see with normal optical apparatus are visible as much larger "shadows." (Picture shows molecule as if photographed from above.)

Into this picture stepped James Watson. He had avoided studying chemistry because he was "too lazy" (which gives hope to many of us). He had previously been especially interested in birds. In graduate school he had attempted to warm benzene over an open flame and had caused an explosion. His advisers concluded that he had no flair for chemistry and forced no further courses on him.

Watson studied for his Ph.D. at Harvard under the famous virologist Salvador Luria, who thought that viruses were "naked genes." About half the mass of a virus is protein and half is DNA, so that the question of which of the two types of molecules actually brings about heredity remained unanswered.

Luria thought that his student Watson would benefit from studying DNA, but only one man in the world was interested in both DNA and genetics at the time, the Danish biochemist Herman Kalckar.

Watson was shipped off to Denmark, on a renewable fellowship, to study DNA and its role in genetics in Kalckar's laboratory. When he arrived, Watson found to his dismay that he could not understand Kalckar's English. He felt doomed to waste the year abroad. A conference at a marine station in Naples, Italy, brought relief from the cold of Denmark and an opportunity to meet Maurice Wilkins, the crystallographer who was attempting to photograph the DNA molecule. For the first time Watson was able to see some clear progress toward an understanding of the structure of the DNA molecule. Wilkins showed him an X-ray photograph proving that DNA was crystalline—that is, that it had a regular pattern of atoms which could, Watson hoped, be worked out.

Soon after this, Linus Pauling wrote a paper describing a model of DNA as an alpha-helix. If Pauling's model was correct, the DNA molecule was in the form of a coiled spring, turning to the right (Figure 26–3).

FIGURE 26–3 An alpha-helix.

These almost-simultaneous events in England and the United States excited Watson because when evidence begins to accumulate at a rapid pace, it often means that a problem is nearing solution. Watson was determined to study the chemistry of the molecule despite his lack of an adequate background. Unfortunately, he did not know enough mathematical chemistry to work with the great Linus Pauling and he could not penetrate the reserve of Maurice Wilkins. However, in Cambridge, England, another biochemist, Max Perutz, was working on a related problem and Watson arranged to study in his laboratory, ostensibly to investigate the structure of muscle protein. Perutz, with the aid of X-ray crystallography, had produced in one day a photograph which corroborated Pauling's hypothetical DNA model!

Unfortunately, the Fellowship Board of the National Science Foundation refused

Watson's request to move from Denmark to England because they felt he should continue his studies rather than switch to a whole new field, X-ray crystallography. Only through complicated negotiations by his advisers at Harvard was Watson able to get even reduced financial support—but by this time he was already in England and determined to go on, despite financial stress.

In Perutz's lab, Watson met Francis Crick, a brilliant, talkative man, a convert from physics to the study of large molecules. It was thus a combination of the biological thinking of Watson, the mathematical skill of ex-physicist Crick, and the crystallographic work of Wilkins which was arrayed against the riddle of the structure of DNA.

Crick soon convinced Watson that "Pauling's accomplishment [the discovery of the helical structure of DNA] was a product of common sense, not the result of complicated mathematical reasoning."

Watson describes the fateful decision he and Crick took when they determined to beat "the world's greatest chemist" at his own game and be the first to discover the structure of DNA. At the time, Watson was all of 23 years old! Here are his words:

> The alpha-helix had not been found only by staring at x-ray pictures; the essential trick, instead, was to ask which atoms like to sit next to each other. In place of pencil and paper, the main working tools were a set of molecular models superficially resembling the toys of preschool children.

> We could thus see no reason why we should not solve DNA in the same way. All we had to do was to construct a set of molecular models and begin to play—with luck, the structure would be a helix. Any other type of configuration would be much more complicated. Worrying about complications before ruling out the possibility that the answer was simple would have been damned foolishness. Pauling never got anywhere seeking out messes.

On pages 329 and 331 you will find pictures of the constituents of the DNA molecule as they were known in 1951. It was already understood that DNA consisted of large molecules called nucleotides linked together in a regular way. These molecules consisted of nitrogenous bases of two types. One type was a purine. It comprised two ring-shaped molecules containing nitrogen. The other type was similar, except that it comprised only one ring. This type was called a pyrimidine.

Purine   Pyrimidine

There were two types of purines and two types of pyrimidines which were always found in chemical analyses of DNA. These were the only constituents of the molecule that varied. For they were attached, somehow, to a chain of alternate sugars (deoxyribose) and phosphate (PO_4) molecules that Watson and Crick decided must always be constant, first a sugar, then a phosphate, then a sugar, and so on.

Cut out the molecules on pages 329 and 331 with the scissors from your dissecting kit. Lay the molecules out in groups, purines together, pyrimidines, sugars, and phosphates. These four types of molecules, Watson and Crick felt, held the secret of heredity.

Now lay out on your desk several different kinds of arrangements of the four kinds of molecules, assuming, as did Watson and Crick, that the sugars and phosphates are a backbone and the purines and pyrimidines somehow come off the backbone. Think of as many variations as you can, in order to get the feeling of the magnitude of the task tackled by those two audacious young men.

Lay out your molecules now! Discuss with your neighbor the variety of options available.

Are there many or few? _____

When you finish, put your molecules aside neatly and read on.

A paper by the biochemist Alexander Todd suggested that the molecule was a helix with the sequence shown in Figure 26-4. This was exciting news because, at last, here was a clear-cut sequence of molecules which might have been close to the answer. But would the data obtained by chemists stand up to the ultimate proof, an X-ray photograph of the molecule itself? Watson and Crick examined a new photograph taken by Rosalind Franklin of Maurice Wilkins' laboratory. It did not conform to Todd's proposed plan, since the molecule photographed was thicker than an ordinary helix (see Figure 26-5).

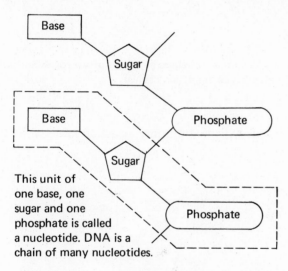

This unit of one base, one sugar and one phosphate is called a nucleotide. DNA is a chain of many nucleotides.

FIGURE 26-4 Sequence suggested by A. Todd.

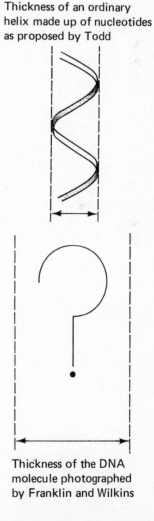

Thickness of an ordinary helix made up of nucleotides as proposed by Todd

Thickness of the DNA molecule photographed by Franklin and Wilkins

FIGURE 26-5 A comparison between Todd's hypothesized DNA helix and Franklin and Wilkins' photograph.

Wilkins suggested that the DNA was a compound helix composed of several nucleotide chains twisted about each other. Although such an arrangement would account for the thickness of the molecule depicted in the photograph, it created another problem for Watson and Crick: before they could employ their proposed technique of model building (trying to imagine how Tinker Toy models of the four kinds of molecules would fit together) they would have to decide how the several helices were bound together. Were they bound together by hydrogen bonds (Figure 26-6)? Or by salt linkages involving negatively charged phosphate

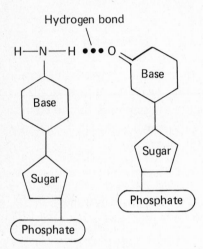

FIGURE 26-6 Bonding of helices by means of hydrogen bonds. (Hypothetical arrangement of molecules)

groups (Figure 26-7)? These were two kinds of bonds which commonly attached molecules of this type together.

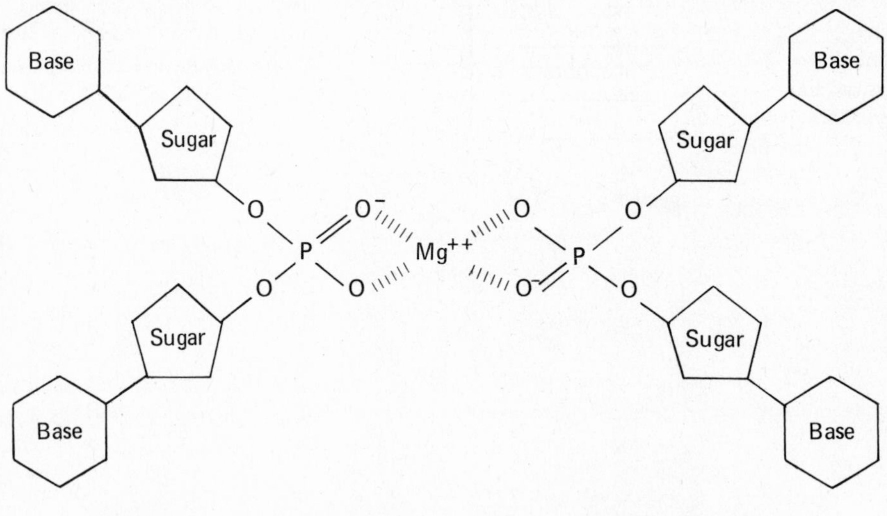

FIGURE 26-7 Bonding of helices by means of magnesium salt linkages. (Hypothetical arrangement of molecules)

The next problem arose from the fact that there are four types of nucleotides found in DNA. Thus DNA was really not a perfectly regular molecule; its nucleotides did not always repeat in the same way. Instead, the four types of nucleotides could appear in any sequence. But the nucleotides were not completely different. They all contained exactly the same phosphate and sugar molecules. They differed only in their nitrogenous bases, and there were only four kinds of nitrogenous base—two kinds of purine (adenine and guanine) and two kinds of pyrimidine (thymine and cytosine). Every nucleotide in DNA has one of these four kinds of nitrogenous bases:

PURINES   PYRIMIDINES

Adenine   Guanine   Thymine   Cytosine

Thus, if the Watson and Crick idea of a backbone of sugars and phosphates was correct, the only diversity in the DNA molecule came from the variations in the four nitrogenous bases. For example, one DNA molecule could have a sequence of ATTAGCAT, while another could have a sequence like GCGCATCCCG. But how could molecules which varied in only four components contain a plan for the construction of a human being or a rose?

The next break came when Crick and a colleague, Bill Cochran, proved that the alpha-helix proposed by Linus Pauling was compatible with Wilkins' X-ray photograph. This was ample proof that the helical structure was the correct beginning step. Then Watson and Crick went to Wilkins' laboratory to look at a new photograph (see Figure 26–8). Can you suggest an interpretation? Each dark shadow is either a nitrogenous base, a sugar, or a phosphate. Your interpretation does not have to be correct. Give it a try by labeling each dot in Figure 26–8 as PO_4, sugar, purine, or pyrimidine.

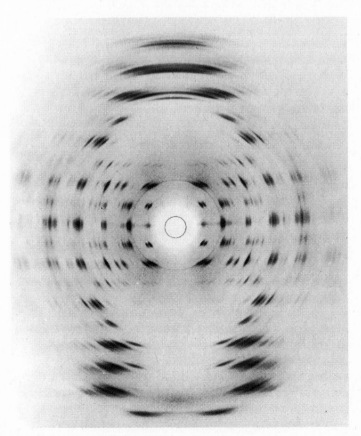

FIGURE 26-8. **The photograph taken by Rosalind Franklin of what she called DNA.** (Picture shows molecule as if photographed from above)

Wilkins felt that his evidence suggested that the helix was constructed of three chains. Crick, upon examining the measurements between the molecules on the X-ray photograph, calculated that the number of chains in the molecule could be only two, three, or four.

BUILDING THE DNA MODEL

The time had arrived for model building. Watson and Crick felt that enough data had been accumulated to try to visualize the DNA molecule. Rosalind Franklin was adamantly opposed to the daring idea of almost arbitrarily putting together more or less imaginary models of an unknown, unseen, and complicated molecule with the hope that its structure would become apparent as various combinations of molecules were moved around in relation to each other. She felt that continued X-ray crystallography and biochemistry would eventually add enough evidence to systematically put the molecule together without excesses of imagination.

1. Give your opinion of the pros and cons of each approach. _____

_____

_____

_____

_____

Watson and Crick began to manipulate their models to examine the following hypotheses

☐ Hypothesis 1: *The molecule consisted of two outside chains of bases joined together by a central skeleton of sugars and phosphates attached by the attraction of magnesium ions* (Mg^{++}).

(Magnesium ions are commonly found in this type of role. See Figure 26–9.)

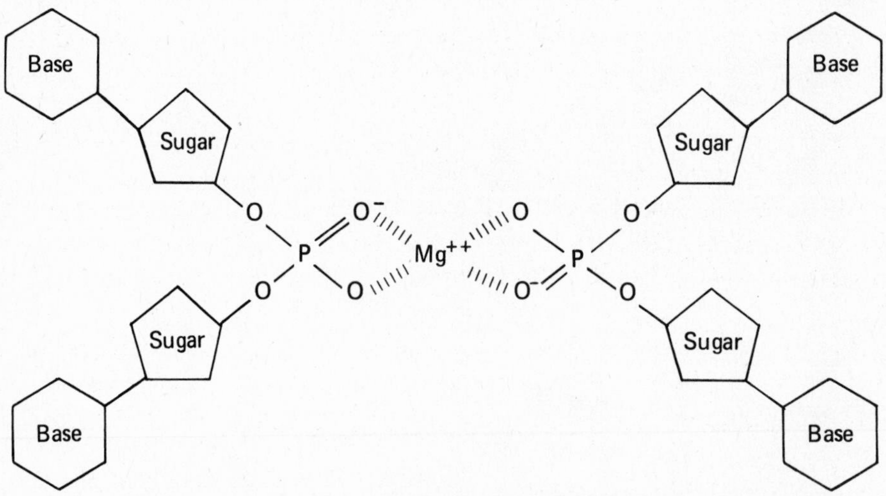

FIGURE 26–9 Two helices tied together by a magnesium ion bond.

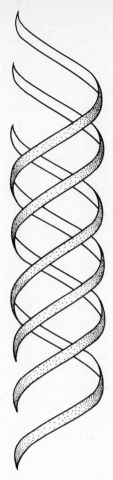

FIGURE 26-10 A hypothetical DNA molecule consisting of three chains twisted around each other.

☐ **Hypothesis 2:** *The molecule consisted of three chains twisted around each other. (See Figure 26-10.)*

Further analysis would have to wait until Wilkins' group had checked out the first hypothesis that magnesium ions were holding the chains together. This shows that theoretical scientists may lack the expertise and equipment to check their own hypotheses. This type of division of labor is becoming common in modern science, where some people devote their time to thinking up possible solutions while other scientists, no less creative or expert, attempt to determine whether or not the theoreticians are correct. Later on we will see that Wilkins shared a Nobel Prize with Watson and Crick in recognition of the importance of the role his data played in the ultimate discovery, and Rosalind Franklin received a touching tribute to her scientific capabilities in the closing pages of Watson's fascinating book.

During this period of waiting for data from Wilkins' laboratory, a new threat appeared on the horizon. Archrival Linus Pauling was on his way to England. If he were to see Wilkins' photograph, he might have the maturity and expertise to solve the problem in short order.

2. Do you think it would have been a good idea for Watson to ask Wilkins not to show his

photograph to Pauling?_____ Explain. _____

_____

_____

Fortunately for Watson, Pauling had previously expressed some politically unpopular views. He was one of the sponsors of a World Peace Conference, and for this the State Department had taken away his passport. This occurred during the McCarthy era, a period of anti-Communist witch-hunting which almost brought an intellectual Dark Age to the United States. Watson wrote, "The reaction was one of almost complete disbelief . . . . The failure to let one of the world's leading scientists attend a completely non-political meeting would have been expected from the Russians." Watson was not as surprised as most, for his original mentor, the famous virologist Salvador Luria, had for similar reasons been refused a passport to go to London to lecture on "The Nature of Viral Multiplication."

3. Comment on the relationship between politics and science. _____

_____

_____

The field was now clear in England. Wilkins' X-ray photographs would not be sufficiently perfected for publication for several months, so Watson and Crick this time had the edge on Pauling.

New evidence came in:

Fact 1: Hershey and Chase, American virologists, sent a letter to Watson indicating that when a certain type of virus (called bacteriophage or phage for short) attacks a bacterium it injects DNA into the bacterium while the protein component of the virus remains outside (Figure 26–11a). Once inside, the DNA causes the bacterium to manufacture many small phage viruses (Figure 26–11b). Thus it was proved that the DNA, not the protein, contained the directions for manufacturing the viruses. *The DNA constituted the viral genes.*

Fact 2: Watson himself, after learning the rudiments of X-ray crystallography, used the facilities in his host's laboratory to take a picture of a related molecule, RNA, and found that it too was a helix.

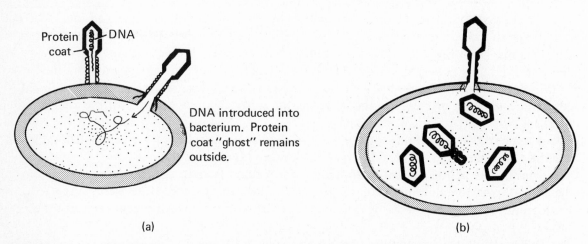

(a)   (b)

FIGURE 26-11 (a) Phage viruses attacking a bacterium. (b) Bacterium manufacturing new phage viruses under the influence of viral DNA. Protein coat of infecting virus still attached to outside bacterial cell wall.

Fact 3: At Columbia University, the Austrian-born biochemist Erwin Chargaff had been analyzing many samples of DNA, and he reported that the number of adenine molecules in all kinds of DNA was the same as the number of thymine molecules. Furthermore, the number of guanine molecules was the same as the number of cytosine molecules. Some organisms had more A and T, and some more G and C; but the proportion of these molecules was always consistent from species to species.

Crick began to think about how genes are reduplicated so that the offspring has the same kind of genetic plan in its cells as the parent. A friend of Crick's, the biochemist John Griffith, calculated that adenine and thymine should stick together by their flat surfaces. Then Crick remembered that these were the bases that Chargaff had found to be present in equal amounts in DNA. This started a new train of thought, since the magnesium ion idea for the attachment of the molecules was no longer the favorite. From this information, Watson conjectured:

> Some fuzzy evidence with sea urchins suggested that DNA was the template upon which RNA chains were made. In turn, RNA chains were likely templates for protein synthesis. Most important point was that DNA, once synthesized, was very stable. The idea of the genes being immortal smelled right.

Then came disaster: Linus Pauling, in February 1953, wrote an article describing the DNA molecule. It was a three-chain helix with a sugar–phosphate backbone in the center. Watson and Crick frantically read a copy of the article. They noticed that Pauling had bound his hydrogen atoms into molecules in a manner which would not permit their release as ions. But DNA stands for deoxyribonucleic *acid*. How could DNA be an acid if it did not liberate hydrogen ions, for the definition of acid is "an excess of hydrogen ions"? Watson and Crick knew that Pauling was wrong; his molecule would not behave like an acid. Real DNA has definite acidic properties.

In late 1952, Rosalind Franklin produced the clearest and simplest X-ray photo of a DNA molecule (see Figure 26–12).

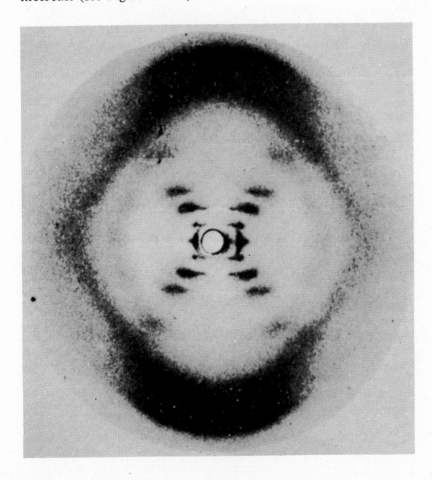

FIGURE 26-12 The 1952 photo of "B" DNA.

The black cross of reflections which dominated the picture could only arise from a helical structure. What remained now was to measure the distances between the first and second layers of lines to find out the number of chains in the DNA molecule. The measurements came out as follows: The helix diameter was about 20 Å (.0000020 mm), and the purine and prymidine bases were 3.4 Å thick and stacked on top of each other in a direction perpendicular to the helical axis.

The information was coming in thick and fast. It was again time to build models. The first model looked like Figure 26–13. It had a phosphate–sugar backbone in the center, with appendages of purines and pyrimidines. But there was too much that did not check out with this model.

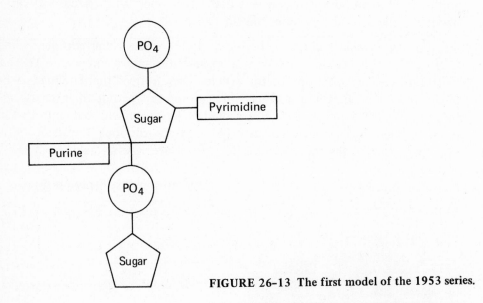

FIGURE 26–13 The first model of the 1953 series.

Then Watson began to consider a double helix, a model with two outside backbones with a repeat of 34 Å representing the distance along the helical axis required for complete rotation (Figure 26–14). He wrote: "Francis [Crick] would have to agree. Even though he was a physicist, he knew that important biological objects came in pairs."

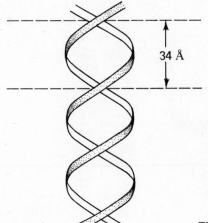

FIGURE 26–14 The double helix.

Watson's perspective now changed. He began to think about how perfectly balanced the X-ray photograph was. How could the purines and pyrimidines be so arranged as to be perfectly balanced? By reading papers by the biochemists J. M. Gulland and D. O. Jordan, he guessed that the bases were attached to one another by hydrogen bonds. The result of this conjecture was the model depicted in Figure 26–15. But the regular appearance of the X-ray photographs would not be duplicated by this model, for purines have two rings and pyrimidines only one. If two purines were joined they would take up more space than two pyrimidines (Figure 26–16). This would make the model in Figure 26–16 look more like the drawing in Figure 26–17.

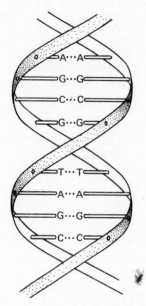

FIGURE 26–15 A model of double helix joined by hydrogen bonds between nitrogen bases.

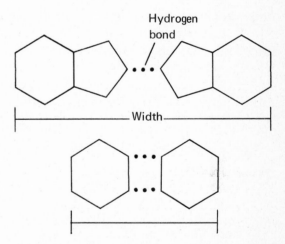

FIGURE 26–16 A comparison between the widths of two pyrimidines compared to the widths of two purines.

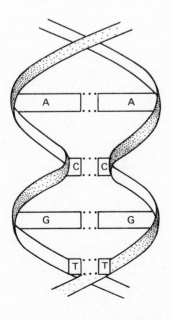

FIGURE 26–17 The appearance of a double helix if pairs of purines and pairs of pyrimidines were attached together.

Furthermore, there were other flaws. This model was incompatible with Chargaff's data which indicated that the amount of thymine equals the amount of adenine and the amount of guanine equals the amount of cytosine. If, theoretically, an organism could have the sequence

A–A
C–C
A–A
G–G
C–C

obviously the amount of thymine does not equal the amount of adenine. In fact, there is no thymine, nor does there appear to be any reason to have thymine present.

Rumor had it that Pauling was coming out with a new model. Time was running out. The shop at Cambridge was to take a week to produce new tin models of molecules for Watson. In desperation he cut out some cardboard models. He shuffled them around and noticed that an adenine–thymine pair was identical in shape and width to a guanine–cytosine pair. This was clearly the key to the problem! It immediately explained Chargaff's data. There were equal amounts of thymine and adenine because they were bound to one another! It also allowed the shape of the outer backbone to be symmetrical because one purine and one pyrimidine bound together took up the same space as the other purine–pyrimidine combination.

Most important, by matching a specific purine with a specific pyrimidine, a genetic code could exist. **For every adenine on one side of the chain, a thymine would be found on the other. For every guanine only a cytosine could be its mate.**

Watson was struck with the beauty of this model. He knew, as did Crick after quick inspection, that this was the answer.

Suppose that the double helix were to split into two single strands during cell division. Fill in the appropriate letters of the single strand which the molecule on the left below would cause to be produced by the nucleus of the cell:

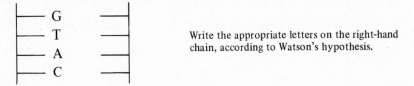

Write the appropriate letters on the right-hand chain, according to Watson's hypothesis.

Draw Watson and Crick's double-stranded molecule in a simplified version which looks like a ladder. Indicate what molecules the two vertical bars are made of. Put in your nitrogenous bases in any sequence, but make sure that the appropriate purine goes with each pyrimidine. Which molecule does the base attach to, the sugar or the phosphate?

Use these molecules in your drawing.

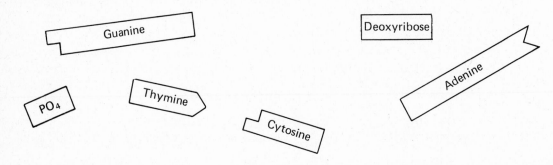

When you have finished your drawing, check it with your partner's. If there are inconsistencies between the two drawings which you do not understand, ask the instructor.

THE MECHANISM OF GENE ACTION

To complete the picture we must now ask, "so what?" Now that we know about the structure of DNA, what does that have to do with my having brown eyes? The answer is that brown eyes (color, shape, formation of eyeball, etc.) are all determined by reactions activated by enzymes (*which are proteins*), and that the eye itself is made of cells, the major constituent of which is protein. In short, to understand how the genetic message is converted into an organism, we must understand how DNA causes the production of proteins. Furthermore, it should be realized that enzyme systems catalyze thousands of reactions inside our bodies every day. These reactions supply us with energy and regulate the other processes which make up life. We have only one nose, two ears, etc. Added up, our physical features are relatively few compared to the enzyme systems inside our bodies which keep us alive. The more important problem, then, is how do our children's bodies get the directions to manufacture the thousands of enzymes they will need? The answer is that sequences of three nucleotides in a row are blueprints for the production of one amino acid, and rows of amino acids comprise enzymes. *Chains of many three-nucleotide units, then, are the closest thing to that mythical concept, the gene.*

Table 26-1 shows an early interpretation of the role of the nucleotides in the production of proteins. On the left are the names of the twenty amino acids which comprise virtually all proteins on the Earth. On the right are the combinations of three nucleotides (designated by their variable constituents, their nitrogenous bases).

TABLE 26-1. The Sequences of Three Nucleotides Representing Each Amino Acid *

Amino Acid	Code of Bases in DNA **
Alanine	GGC
Arginine	GCG
Asparagine	TGT
Aspartic acid	TGT
Cysteine	AAC
Glutamic acid	TGT
Glycine	ACC
Histidine	TGG
Isoleucine	AAT
Leucine	CAA
Lysine	TTT
Methionine	ACT
Phenylalanine	AAA
Proline	GGG
Serine	AGC
Threonine	GTG
Tryptophan	ACC
Tryosine	ATA
Valine	ACA
Glutamine	GTC

*Modified from John Abelson, *Science,* 139, p. 775 (1963).

**These sequences were published by a National Institutes of Health team. They have since been modified.

■ THE PROBLEM: *To build a model of a DNA molecule which will cause the production of a specific amino acid chain.*

Pick one of the three nucleotide sequences which represent an amino acid from the "Code of Bases" column of Table 26-1. Specify which amino acid you have chosen.

_____ What is the sequence of three nitrogenous bases which represents

that particular amino acid? _____   _____   _____

Select those bases and their counterparts (for example, a G for every C) from the molecules you cut out at the beginning of this exercise.* Tape each molecule to the next at the appropriate tabs. Always have a hydrogen (H) showing at the taped bond between bases to indicate a hydrogen bond. Each dotted bond on a nitrogenous base joins with a hydrogen on the complementary base. The bonds to the phosphate are indicated by the symbol PO_4. The sugar–base bond is indicated by an arrow. Notice that some of the molecules will fit on the

*If this is one of the last laboratory periods of the academic year, you can use the models printed inside the covers of this manual. The heavier paper will make a more satisfying helix.

right-hand side of the "ladder" and some on the left. **It will be necessary for you to turn some of the molecules over.** It will be easier if you build your model by first attaching the nitrogenous bases together in the sequence you want and then attaching the sugar–phosphate backbones.

When you have completed your molecule correctly, tape it to the sequence built by your neighbor, so that you have six nucleotides in the chain. Then add another neighbor's contribution until at least four sets of three nucleotides each are taped together. Let the molecule hang over the edge of the table. Use a ring stand as the central core of your molecule. Tape the molecule to the stand at appropriate intervals so the molecule keeps its shape as a helix.

4. How many nucleotides are needed for one complete turn of the helix? _____
To calculate this turn to page 322.

When you have finished your massive DNA molecule, name the amino acid chain which it will dictate. The name will consist of the individual names of the component amino acids. *Example:* The sequence AAT CAA TTT ACT would cause the production of isoleucyl-leucyl-lysyl-methionic acid.

Name of your molecule: _____

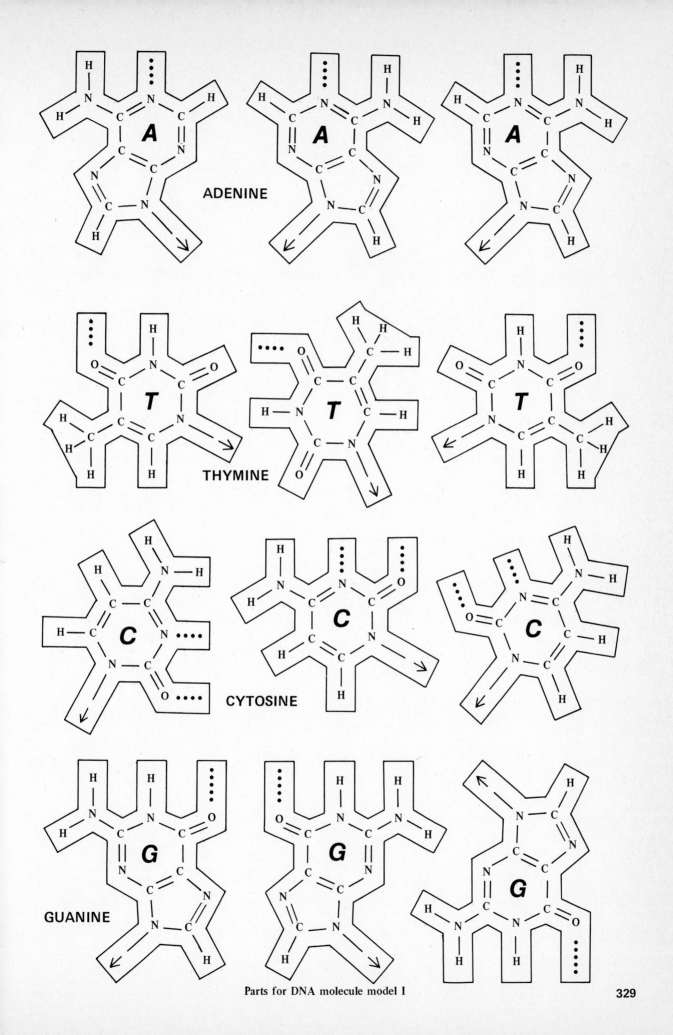

Parts for DNA molecule model I

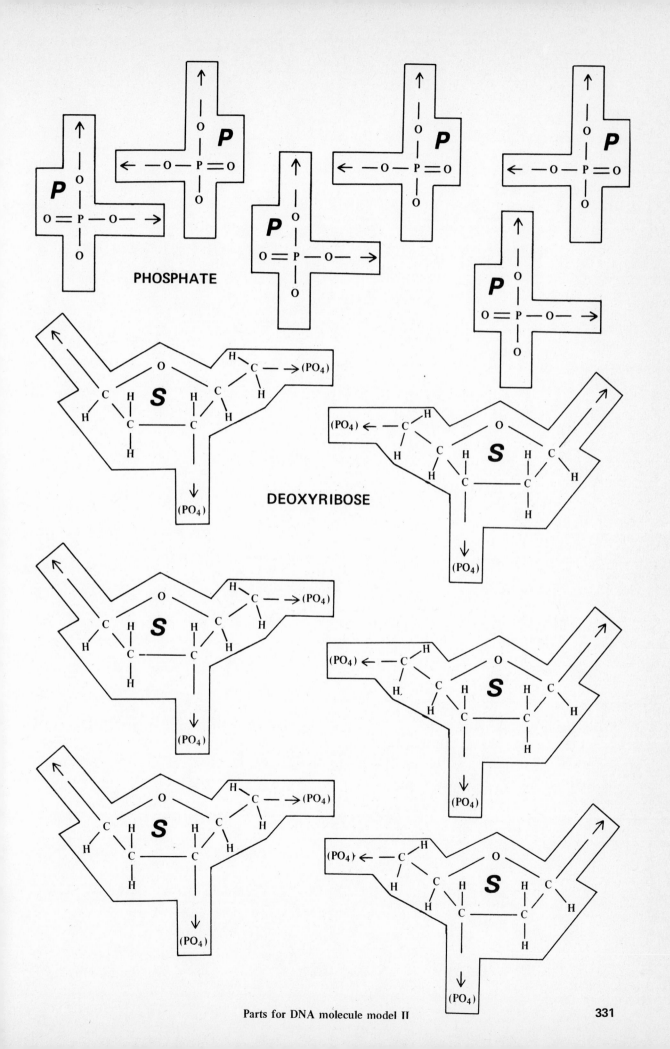

PHOSPHATE

DEOXYRIBOSE

ADDENDUM: TRANSCRIPTION AND TRANSLATION

By Kathleen Sacco

The genetic code you discovered while making the DNA molecule is the basis for what is called a gene.

Three consecutive bases on the DNA molecule "code for" a specific amino acid (for example, GGC is the code for alanine). The chain of sugars and phosphates together with three consecutive bases is called a **codon**; it codes for one amino acid. Imagine a long DNA chain, consisting of 300 pairs of bases. This represents 100 codons, or the ability to code for 100 amino acids. (There are only 20 different amino acids in proteins, but they can be arranged in an almost infinite number of sequences.) When these 100 amino acids are bonded together, they form a certain protein, X. The linear arrangement of nucleotides determines the linear sequence of amino acids in a protein (polypeptide). The segment of a DNA chain which produces one polypeptide is called a "gene." How does all this happen?

An enzyme molecule called RNA polymerase "reads" the DNA chain and creates a **messenger RNA (mRNA)** strand in a process called **transcription**. The polymerase works like a tab on a zipper; it splits the DNA into two halves, snipping the hydrogen bonds between the base pairs as it moves down the strand. The DNA is bathed in a pool of floating nucleotides (a nucleotide consists of one sugar, one phosphate, and one base). As the polymerase disconnects each base pair, it picks up the appropriate nucleotide to pair with the disconnected base. The polymerase reads only one side of the DNA chain; it does not skip back and forth. As more bases are disconnected, more new nucleotides are added, and the mRNA gets longer. After the polymerase "reads" the DNA and passes on, the DNA base pairs rejoin as a double helix. **When the whole message is read, the DNA is unchanged and a new mRNA exists.** A message can be for one or more genes.

5. What do you hypothesize to be the role of the mRNA? _____

_____

_____

Procedure

Cut out the DNA chain, tRNA's, amino acids, and nucleotides on pages 337, 339, and 341. With your scissors acting as the RNA polymerase, snip apart the first pair of bases. Reading the strand on the left, pick the corresponding base from the nucleotide pool. **U (uracil) takes the place of T (thymine) in mRNA.** As you read the DNA strand, join the mRNA nucleotides to each other with tape. When you have cut your fourth base pair, rejoin the first pair with tape; continue this process like two zippers following each other (three base pairs apart), one opening, the other closing the DNA helix. Remember *not* to attach the mRNA to the DNA.

TABLE 26-2 Amino Acids and Their Codons*

Phenylalanine	UUU, UUC
Isoleucine	AUU, AUC, AUA
Valine	GUU, GUC, GUA, GUG
Proline	CCU, CCC, CCA, CCG
Alanine	GCU, GCC, GCA, GCG
Histidine	CAU, CAC
Asparagine	AAU, AAC
Aspartic acid	GAU, GAC
Cysteine	UGU, UGC
Arginine	CGU, CGC, CGA, CGG, AGA, AGG
Leucine	UUA, UUG, CUU, CUC, CUA, CUG
Methionine	AUG
Serine	UCU, UCC, UCA, UCG, AGU, AGC
Threonine	ACU, ACC, ACA, ACG
Tyrosine	UAU, UAC
Glutamine	CAA, CAG
Lysine	AAA, AAG
Glutamic acid	GAA, GAG
Tryptophan	UGG
Glycine	GGU, GGC, GGA, GGG

*Notice that there are 61 different combinations of the four letters C, U, G, and A listed above. If you try to visualize a chain of 100 codons, you will begin to understand why there are almost an infinite number of combinations which can comprise just one DNA molecule. And there are thousands of DNA molecules in the nucleus of one cell.

Notice that the mRNA begins with AUG. This is always the beginning codon of a message; it results from the fact that the first codon of every message on the RNA is TAC and it tells the polymerase where to begin reading the DNA. After completing your mRNA, determine its amino acid sequence by checking the amino acid–codon list in Table 26-2. In the same manner that AUG begins the message, UAG, UAA, or UGA ends the message. You will discover why when you construct an amino acid chain (polypeptide). When you are through, the DNA chain (gene) should be unchanged, and, in addition, you should have an mRNA transcribed from that gene.

The messenger RNA now brings its transcribed sequence of nucleotides out of the nucleus to the **ribosomes**, spherical structures found on membranes scattered throughout the cytoplasm. **Translation** (reading an mRNA and substituting the proper amino acids) occurs on the ribosomes.

Gather together your mRNA, transfer RNA's (tRNA), and amino acids. Fit together a tRNA with its proper amino acid. They are now ready to move together with the tRNA acting as a tow truck. In the cell the tRNA picks up the appropriate amino acid in the cytoplasm and brings it to the ribosome, where it will be added to a chain of amino acids to make up a polypeptide (protein).

Imagine that you are a ribosome beginning to read the mRNA. (A ribosome normally forms two identical polypeptide chains simultaneously; we will form only one.) Notice that each tRNA has a specific code at the anticodon end which is opposite the amino acid attachment end. The anticodon and codon bases "fit together" and will naturally pair. Fit in the first two tRNA's with their amino acids. Notice that the amino acids are lined up to make a C–N bond at the arrows. Tape this bond. These *two* bonded amino acids form a *di*peptide. Free the first tRNA and bring in the next one. Attach the next amino acid to the previous one. Continue this process, remembering that only two tRNA's can be "translated" at one time. When you are finished several amino acids will be bonded together as a *poly*peptide. The tRNA's will be used again when the amino acid they attach to is needed.

   6. What happens when the last codon is read? _____

_____

_____

   7. Is there a corresponding tRNA and amino acid to the last codon according to Table 26–2? _____

_____

   At this point, when the last RNA leaves, the protein is completed and will fall off the ribosome.

Mutations

   Most proteins consist of hundreds of amino acids. The hemoglobin molecule consists of two strands each of two different proteins (two A's and two B's). A and B are coded for by two different genes. The whole molecule is 582 amino acids long! If one incorrect amino acid is placed into each of the two B strands by a malfunction of the B gene, the molecule will be imperfect. Sickle cell anemia is caused by such a defect in one amino acid! In the course of evolution, a genetic mutation occurred that changed one codon in gene B, causing it to call for the wrong amino acid.

   There are many different types of mutations. Take your mRNA strand and, following the instructions, cause it to mutate. *Remember that DNA mutates, not RNA!* However, it is too time consuming to mutate the DNA chain and then construct a new RNA each time.

   The most common mutation is **deletion**, the removal of a base or, less frequently, of a whole codon. Remove the fourth base from your mRNA and join the two cut surfaces. Now read the codons in sequence and discover what amino acids are called for.

   8. What has happened to all the codons following base No. 4? _____

   Next, return the fourth base to its proper position. At the same time, add a new nucleotide (any one) between the original fourth and fifth bases. This is an **addition**; it occurs infrequently.

   9. Now what happens to all the codons beyond this new nucleotide? _____

_____

_____

10. What will happen to the amino acid chain? _____

_____

_____

Return the mRNA to normal again. We will now mutate the RNA by **substitution**, adding the wrong base to the mRNA strand. There are two types of substitution. First change base No. 4 from G to C. The second codon now reads CCG, and the wrong amino acid will be brought in by a tRNA. However, all the following codons remain correct. Again, return the chain to normal. Now substitute a U for an A in base No. 10.

11. What will be the effect of this new codon? _____

_____

_____

12. What would happen to a polypeptide consisting of 59 amino acids which had had this

change at amino acid No. 20? _____

_____

_____

13. From this exercise do you feel that most mutations are beneficial or harmful?

_____ Explain in detail below.

Part of a "gene" consisting of a number of attached nucleotides.

Cut out both chains. Tape top of second chain (above) to bottom of first.

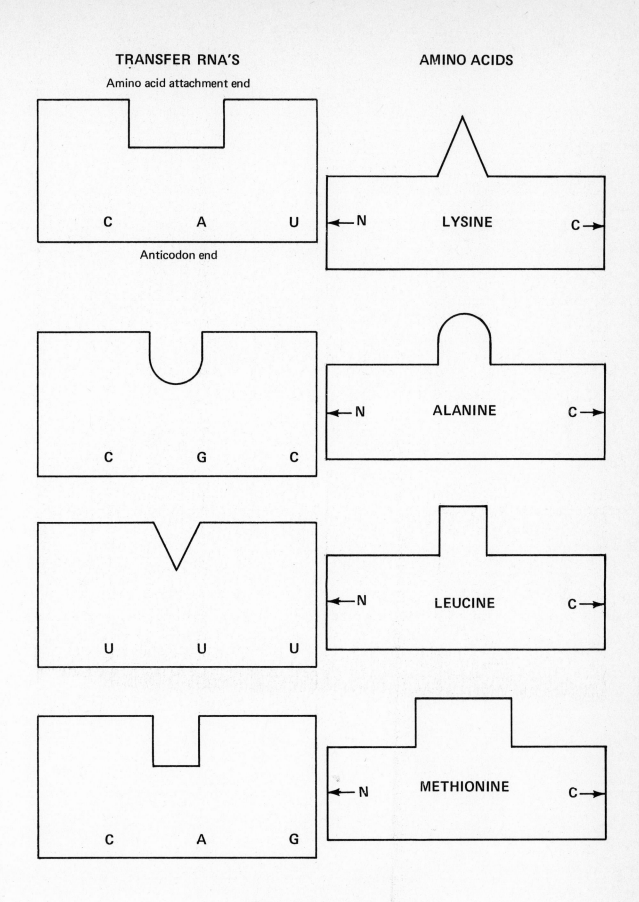

Transfer RNA's and amino acids.

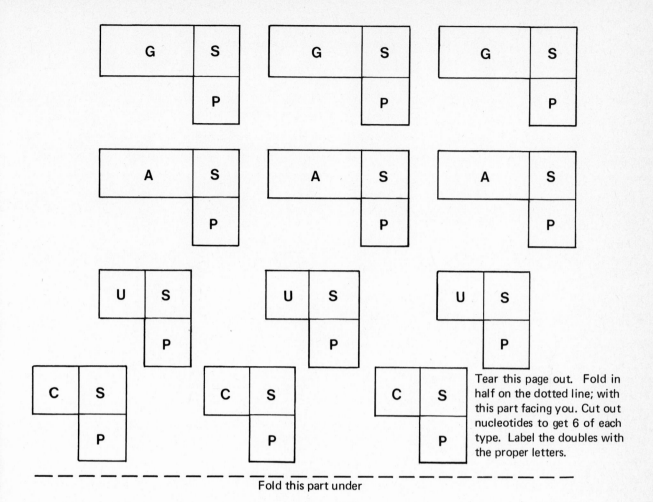

Tear this page out. Fold in half on the dotted line; with this part facing you. Cut out nucleotides to get 6 of each type. Label the doubles with the proper letters.

- -
Fold this part under

Nucleotide pool (free nucleotides in the nucleus; when attached, they will constitute an RNA molecule).

IX Animal Behavior

27 Anthropomorphism

Kinesis and Taxis

Before reading on, fill in Table 27–1. *Place a check next to each kind of animal on the list which you feel is capable of dreaming while asleep.*

TABLE 27–1

Pigeons _____	Parakeets _____	Goldfish _____	Wolves _____
Earthworms _____	Snakes _____	Snails _____	Chickens _____
Guppies _____	Lions _____	Dogs _____	Clams _____
Cats _____	Giraffes _____	Alligators _____	Monkeys _____
Cockroaches _____	Grasshoppers _____	Crabs _____	Humans _____

1. Which organisms are you sure are capable of dreaming? _____

_____

2. What evidence do you have? _____

_____

_____

Your instructor will tabulate the responses of the class on the chalkboard.

3. Has a pattern appeared? _____ Summarize the attitudes of your class

toward the ability of animals to dream. _____

_____

_____

_____

_____

Dreaming, as far as we can tell, is a human trait, although there is some evidence that dogs exhibit the rapid eye movements (REMs) which characterize human dream periods.

The attribution of human characteristics to animals is called **anthropomorphism**. Usually the closer an organism is to man, either phylogenetically (for example, vertebrates like dogs and cats are closer on the evolutionary scale than insects or clams) or in familiarity (for example, goldfish and parakeets are more familiar to us than carp and orioles), the more likely people are to ascribe human characteristics to it.

Living with animals has always been a part of human experience. Sharing one's home with a flock of goats is a dimly remembered ancestral memory to some of us, although it is still practiced in many parts of the world today. But the luxury eighteen-story apartment in midtown is not free of its complement of animals. The cockroach is found wherever man lives. The ancient silverfish, a primitive insect which has existed unchanged for millions of years, can live as comfortably on the glue of bookbindings as it did on scraps from the caveman's table or on rotting plants in the moist fern forests which existed before vertebrates evolved.

Attitudes toward animals vary with human experience. Some people are offended at the idea of placing stones in the uterus of female camels being prepared for a long caravan trip across the desert, a form of birth control practiced by Bedouin tribesmen. The same people may be less offended when an offshoot of this practice, the IUD (intrauterine device) or "coil," is placed in a woman's uterus for the same purpose—to prevent conception.

What are the behavioral limitations of nonhuman animals? The cat, the crocodile, and the scarab beetle were worshipped by the ancient Egyptians as gods capable of superhuman feats. The same beetles in our culture are often squashed on sight. Do insects have "feelings"? Do they, for example, feel pain? The child brought up on a farm is accustomed to the sight of animals being slaughtered for food, and he finds it relatively easy to participate in preparing a hog for market. His city cousin is horrified at the idea, yet he squashes every spider and insect in his apartment, and the recognized method of removal of mice is to flush them down the toilet.

The emotional response to animals (genuine love for dogs and cats; abhorrence of snakes and spiders) can cloud rational behavior. Antivivisectionists describe bloodcurdling experiments supposedly performed in hospitals and laboratories on dogs and cats. To attribute excessive cruelty to animals to scientists is to deny the scientists the humanity they are entitled to, just as to destroy an animal because it offends one's esthetic sense deprives that animal of its inherent dignity as a living thing.

4. Does man have the right to use animals in place of humans for legitimate experiments which might help improve human health or well-being? _____ Explain.

How "human" are the lower animals? Do they think? Is their behavior governed by emotions? The octopus, a close relative of the clam (both are molluscs), clearly exhibits emotions. We can tell, because it "blushes." It changes color rapidly with its "mood," and since the colors are associated with different "states of mind" (a dull color when at rest, rapid color changes to purple and pink when disturbed), we can relate these color changes to its behavior.

Today's exercise will give you a chance to test the level of behavior of another relative of the octopus, the snail, and a species which has survived for millions of years before man appeared, the isopod.

PRELIMINARY INFORMATION

An **ecological niche** is that part of the environment which includes food, shelter, and other factors necessary to sustain an organism. An unusually complex and well-inhabited ecological niche is found under the surface of a rotting log in the forest. Animals and plants living there have become adapted to a narrow set of conditions. Vary the delicate balance even slightly and the organisms will die. The major **limiting factor** is the availability of moisture. Animals which seek dark, moist environments by tunneling through the deteriorating wood live in an atmosphere of 100% humidity. Isopods, important members of the rotting-log community, have evolved from sea-dwelling forms and have never really achieved a satisfactory substitute for gills. They will die if kept in the very low humidity of a heated house or classroom for more than an hour or so. (The relative humidity of your house in the dead of winter with the heating system operating maximally is lower than that normally found in the Sahara Desert.)

The rotting log has its **herbivores** (termites which eat the wood, millipedes which graze on molds which, in turn, obtain their sustenance from the decaying organic matter in the log). Also living in the tunnels and spaces which honeycomb the log are the **carnivores**, especially the centipedes which have a pair of poisonous claws with which to paralyze the small worms, insects, and isopods upon which they prey. Snails, moving along on trails of mucous, are relatively impervious to the centipedes because they can withdraw into their shells. A snail breathes through a small hole in its side, called a pneumostome, which opens into a moist chamber surrounded by blood. This primitive lung is adequate in the moist air of the rotting-log community.

Two highly successful inhabitants of rotting logs are the isopod *Armidillidium vulgare,* often called the sowbug, and land snails of the genus *Polygyra* (Figure 27–1).

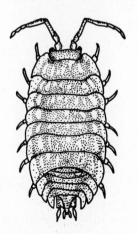

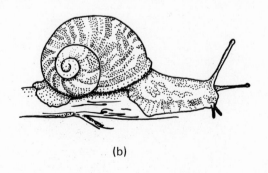

(b)

(a)   FIGURE 27–1   (a) *Armidillidium*, an isopod. (b) *Polygyra*, a land snail.

PART I. OBSERVATIONS

At your table will be found two closed containers, one labeled "Isopods" and one labeled "Snails," and a white enamel tray marked off into four quadrats. Work in pairs, one partner obtaining ten isopods from the container and placing them into a petri dish. The other partner will obtain ten snails in another petri dish, an enamel tray, two discs of filter paper, and a ruler.

Observation 1

Isopods are particularly susceptible to a lack of moisture. To see if their behavior is affected by availability of water, wet the disc of filter paper and place it in the middle of one quadrat. Pour all ten isopods into the center of the tray and record in Table 27–2 what happens after 30 seconds, 1 minute, 3 minutes, and 5 minutes. Repeat the procedure three times, or until you are satisfied that you have accumulated enough data in Table 27–2 to make a generalization.

TABLE 27-2   Isopod Distribution

Time	Number of Isopods in Quadrat I	Number of Isopods in Quadrat II	Number of Isopods in Quadrat III	Number of Isopods in Quadrat IV
30 sec.				
1 min.				
3 min.				
5 min.				

5. Make a generalization about the distribution of isopods in the space below.

6. Are you reasonably sure of this generalization? _____ Explain why or

why not. _____

_____

_____

7. Is there a stimulus for this behavior?   That is, if the isopods seem to move in a
particular direction, hypothesize as to what it is they are moving toward or away from.

□ Hypothesis I: _____

_____

Test your hypothesis. Explain what you have done and the results in the space below.
Keep testing alternative hypotheses, if necessary, until you feel that you have adequately
explained the cause of the behavior of the isopods as far as their distribution is concerned.
Record what you did below.

8. What do you believe to be the cause of the isopods' behavior? _____

_____

_____

_____

Measure the distances between adjacent isopods in the quadrat which has the most isopods
in it. For example, if quadrat II has six isopods in it and the other quadrats have fewer than

six, you would measure the distance of each isopod from its nearest neighbor in quadrat II. Calculate the average distance.

$$\text{Average} = \frac{\Sigma X}{N} \quad \text{or} \quad \frac{\text{sum of all measurements}}{\text{number of measurements}}$$

Average distance between isopods _____ mm

    Put the isopods back into the container. Close the lid.

Observation 2

    Repeat for the snails the procedure for the isopods, using ten snails. Fill in Table 27–3.

TABLE 27–3 Snail Distribution

Time	Number of Snails in Quadrat I	Number of Snails in Quadrat II	Number of Snails in Quadrat III	Number of Snails in Quadrat IV
30 sec.				
1 min.				
3 min.				
5 min.				

    Repeat the procedure three times, or more if necessary.

    9. Is the snail behavior the same as that of the isopods? _____

    10. Make a generalization about snail behavior in the space below.

    Measure the distances between neighboring snails in the most highly populated quadrat.

Calculate the average distance between snails. _____ mm

Summary

Examine all the data you have accumulated and the explanations of behavior you have tested. In the space below, discuss the behavior of isopods and snails, including *factors affecting their distribution.*

PART II. EXPERIMENTS

The isopod *Armidillidium vulgare* exhibits a clear-cut behavior pattern which governs its distribution.

11. Does this pattern have survival value? _____ Describe an instance in

which this behavior might help the isopods survive. _____

_____

_____

_____

Two problems have arisen from the observations we have made. These concern the degree to which the behavior of these animals is consciously controlled. In other words, do the organisms "decide" on the type of response they will make to a stimulus?

There are several categories of behavior exhibited by lower animals. The simplest type is a **kinesis**. This type of behavior is characterized by random movements which become more and more vigorous as the stress from an unfulfilled physiological requirement becomes greater and greater. For example, if an animal is hungry, it will move about rapidly until it finds food (**appetitive phase**). When it finds the food, it engages in a **consummatory act** (eating), triggered by the stimulus of the food itself. The final phase is **quiescence**. The animal slows down its activity related to satisfying the original physiological requirement. It might even go to sleep after its meal, for example.

A kinesis, then, is characterized by **random movement** which continues until a particular physiological condition is reached. Then it stops. The physiological condition may be hunger, thirst, sexual fulfillment, or habitat selection.

A **taxis**, on the other hand, is not random. The organism points itself at (or away from) the stimulus and approaches (or moves away from) it *directly*. When it reaches the origin of the stimulus, it engages in the consummatory act and becomes quiescent. (Or when it gets far enough away so that the stimulus is no longer effective, it becomes quiescent.)

The response may be either positive or negative. The name of the taxis usually has the appropriate Greek prefix attached to it. Thus an attraction toward the light, such as is exhibited by a moth, would be *positive* **phototaxis** (*photos,* the light), movement *toward* light. An organism crawling *upward* in a dark place (that is, not influenced by light) would be exhibiting *negative* **geotaxis** (*geo,* earth), movement *away from* the center of the Earth's gravitational field.

- PROBLEM I: *Is the behavior affecting the distribution of isopods an example of kinesis or of taxis?*

- PROBLEM II: *Is the behavior affecting the distribution of snails an example of kinesis or taxis?*

At the front table you will find a T-maze. Obtain 10 snails and 10 isopods and use the T-maze to solve each problem. On the Report Sheet (page 359) be sure to record the following after you have completed your experiments and analyzed your data.

(1)   Your hypothesis.
(2)   The observations which led to your hypothesis.
(3)   All the information you obtain from testing your hypothesis.
(4)   Your statistical analysis.
(5)   Your experimental and control data.
(6)   The solution to the problem.

Use of the T-Maze (Figure 27-2)

(1)   Obtain a wad of cotton wool, wet it, squeeze out excess water, and place it at one end of the T-maze, against the stopper. Make sure both stoppers are in tightly.

(2)   Unscrew the bottle and place organisms inside. Screw bottle back in tight. Keep bottle vertical so that organisms do not fall out.

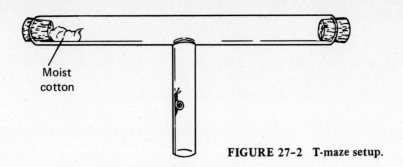

Moist
cotton

FIGURE 27-2 T-maze setup.

(3) Place the maze flat or vertical, whichever seems more appropriate. If the flat placement is selected, the bottle should be supported by a pencil on the table, at right angles to the bottle, to keep it level.

Begin your experiments now. Record your data below. Repeat each experiment at least five times and organize your data in a table.

Analysis of the Data

One of the most discomforting problems facing the scientist is that of determining whether or not his data are significant. In short, if you used 10 snails and 6 went left and 4 went right, could you make any generalization about the response of snails to a possible stimulus on the left? What if 7 went left and 3 went right? If all 10 went left, would you then be sure that the proposed stimulus was the cause of their behavior?

There are just a few basic tests of significance that are usually applied to find out the significance of differences between populations above 30, where chance can operate. One of these is the chi-square test. It has additional usefulness because it is not as dependent as some of the other techniques on sample size and can be used with relatively small samples, such as 10. Unfortunately, the smaller the sample size, the greater must be the difference between groups in order to show significance. In other words, if 7 out of 10 snails were found on the left, it is more likely that this may have been an accident (chance) than if 700 out of 1000 snails turned left. With this understanding, apply the following chi-square test to each of your sets of data to find out whether or not it was significant.

The Chi-square Test

The purpose of this statistical test is to determine whether the frequency of a particular response in your samples is accidental (that is, by chance) or the result of some systematic influence. In order to find this out, you compare the frequency of responses exhibited by your isopods or snails to the expected frequency (chance frequency).

12.   In a situation where the animals could go either left or right, what would you expect

the chance frequency to be (that is, what percent would go left by chance)? _____ %

In order to compare the obtained frequency (your data) with the expected frequency, (chance data) follow this procedure:

(1)   Compute the number of snails you would expect to turn left if chance were operating. Your sample size was probably 10. Of these 10, how many would normally

turn left if there was no particular influence acting on them?_____ How many

would turn right?_____ These numbers are your *expected frequencies.*

(2)   Take a set of data from your table on page 353. Record the number of snails which

turned left_____ and right_____

(3)   Fill in the spaces in the formula below.

$$\chi^2 = \frac{\left(\begin{array}{cc} \text{Number of snails} & \text{Expected} \\ \text{turning left} & \text{frequency} \end{array}\right)^2}{\text{Expected frequency} _____} + \frac{\left(\begin{array}{cc} \text{Number of snails} & \text{Expected} \\ \text{turning right} & \text{frequency} \end{array}\right)^2}{\text{Expected frequency} _____}$$

(4)   Subtract the expected frequency from the actual frequency of left turns and of right turns. If the expected frequency is greater, your answer will have a minus sign in front of it. This will disappear when you square each difference, but you must keep it in mind in order to determine whether or not the difference in frequencies is positive or negative.

(5)   Square each difference and divide the result by the appropriate expected frequency.

Add these numbers together. Record your answer. χ^2 = _____

(6)   To find out what your chi-square number means, find the number *just lower* than yours in Table 27-4.

TABLE 27–4   Chi-square Values at One Degree of Freedom

Level of Confidence	.50	.30	.20	.10	.05	.02	.01
Minimum Value of Chi-square at DF 1	.455	1.074	1.642	2.706	3.841	5.412	6.635

The upper row is the level of your confidence that your results *did not* occur by chance. For example, let us suppose that your chi-square number was 1.73. The number on your chi-square table closest to it (and just below it) would be 1.642. The heading of the column above 1.642 is its level of confidence, .20. This means that you could expect results like this to occur less than 20% of the time by chance. Seems like a good bet? Not good enough. Depending on the nature of the data, scientists set levels of confidence at .05 or .01 *before they begin their experiments*. (.05 means you would expect results like this to occur rarely by chance—less than 5% of the time.)

13. Why is it necessary to set up levels of confidence *before* starting an experiment?

_____

_____

_____

In any field of investigation, including psychology, education, physics, etc., a level of confidence is always indicated, the .05 level usually being the lowest acceptable. It means that the results of the experiment (obtained frequency) differ from the expected frequency (chance frequency) so that only 5 times out of 100 would the event be expected to occur by accident (chance).

We will use the .05 level of confidence for our investigation.

Calculate the chi-square of each experimental trial you have performed.

TABLE 27-5   Comparison of Obtained Chi-squares with .05 Level of Confidence

	I Chi-square Number	II .05 Level of Confidence	III Accept or Reject Null Hypothesis
Isopods			
Trial 1			
Trial 2			
Trial 3			
Trial 4			
Trial 5			
Snails			
Trial 1			
Trial 2			
Trial 3			
Trial 4			
Trial 5			

Record the chi-square number you calculated in column I of Table 27-5. Record the .05 level of confidence number found on the chi-square table in column II. Is your chi-square number higher? If so, write "reject" in column III. This means you *reject the null hypothesis.* The null hypothesis is "There is no significant difference between the experimental and control variables." Or, in our investigation, "There is no significant difference between the expected frequency and the obtained frequency." To reject the null hypothesis means that you *do not* believe that the results are caused by chance so they *are* the result of some systematic factor in your experiment.

Now it is finally possible to decide whether or not the snails and isopods we studied exhibited kinesis or taxis in their distribution patterns.

14. Do we have any information regarding the presence of nonrandom factors affecting the distribution of snails and isopods? Indicate why you do or do not believe in the existence of a factor or factors which affect the distribution of

a. Isopods _____

_____

_____

b. Land snails _____

_____

15. If there was any systematic behavior on the part of the snails or isopods, it may have been a kinesis or a taxis. How could you tell which it was? _____

_____

_____

_____

(If you could not answer that question, go back to page 352 and reread the definitions of kinesis and taxis.)

Conclusion   (Also include the answers to these questions on your report sheet.)

16. Is the behavior affecting the distribution of isopods an example of kinesis or taxis? _____ Explain. _____

_____

_____

17. How sure are you? (Were most of your chi-squares above the .05 level of confidence?)

_____

18. Is the behavior affecting the distribution of snails an example of kinesis or taxis? _____ Explain. _____

_____

_____

19. How sure are you? (Were most of your chi-squares above the .05 level of confidence?)

_____

20. Someone said that if you reject the null hypothesis the behavior is a taxis, and if you accept the null hypothesis, it is a kinesis. Is such a statement reasonable? _____

Explain. _____

_____

_____

REPORT SHEET   Name _____

Instructor's name _____

State problems, hypotheses, results, and conclusions. Include evidence and results of statistical analysis.

X Ecology

28 An Independent Investigation by the Student

Effects of Environment on Development of Brine Shrimp Populations

The preceding exercises were devoted to helping you understand the processes by which scientific ideas are developed. Thinking strategies, such as deduction and the development of theories, were separated out of the context of the overall scientific process and clarified. Now it is your turn to put together these strategies to solve a problem of your own. This exercise is an opportunity to apply your training in a nonstructured situation similar to that of the scientist in his laboratory. You will be given the eggs of a small aquatic organism and asked to study the effects of some aspect of the environment on their hatching time/rate. You are to think up the specific problem you will study, and you will provide your own hypotheses, experimental design, and solutions.

PRELIMINARY INFORMATION

The brine shrimp, *Artemia salina*, is preyed upon by an unusually large number of organisms, primarily because it is "bite size." When hatched from its egg, the larva serves as food for young fish, and as the brine shrimp grows to its mature size, about ½ inch, larger fish prey upon it. Since its growth rate coincides with the developing needs of its growing predators and since it is particularly vulnerable to fish larvae — the most abundant stage in the life cycle of the fish — *Artemia* is hard-pressed to survive.

In the course of millions of years of competition for survival in the face of this high rate of predation, *Artemia* has evolved two mechanisms that ensure its existence as a species. One is its high **biotic potential** (ability to produce many offspring). A single brine shrimp can produce as many as several hundred eggs of two types. One type hatches during the summer; the other is resistant to drying and climatic change and overwinters to hatch the next spring. This capability of the egg to resist drying contributes to its second survival mechanism — it can live in an environment so inhospitable as to preclude the presence of most predators. For example, there are colonies of brine shrimp in Great Salt Lake in Utah and in the salt works in San Francisco, where the salinity is so high that it taxes the ability of most other organisms to maintain their osmotic equilibrium. Furthermore, the capacity to produce resistant eggs allows *Artemia* to survive in what are called "temporary" bodies of water. There are ponds and even puddles which may appear occasionally in the desert and then remain dry for years before it rains. Two days after rain, these formerly dry ponds and pools are swarming with brine shrimp!

■ THE PROBLEM: *What are some environmental factors which affect the rate of embryological development or hatching time of Artemia salina?*

Materials

At the front table you will find vials of *Artemia* eggs, salt, distilled water, small petri dishes, thermometers, graduated cylinders, and chemical balances. Space for your dishes in an incubator and/or refrigerator can be obtained by consulting your instructor.

Protocol

Plan out your whole investigation *after reading the rest of this exercise.* After you have every step planned, follow the directions below to begin the egg-hatching procedure.

> *Directions.* Sprinkle approximately 100 eggs (practically the smallest amount you can pick up with the spatula) on the surface of the water in a half-filled petri dish and cover it until it is to be observed. Count the eggs under a binocular dissecting microscope. (**Note:** These directions are to be followed after you have planned out your study and know exactly what you are about to do.)

Make up one or several hypotheses, test it (them), record your data, and arrive at some conclusions. Hand in your report in one week. Your time schedule should be something like this:

> *Day 1 (today).* Go to the library to obtain information about the development of *Artemia* eggs and the factors influencing the rate of egg development.
>
> *Day 2.* Make up hypotheses, plan experiments, gather materials together, and begin experiments. Record appearance of eggs as viewed under binocular dissecting microscope.
>
> *Day 3.* Examine eggs under binocular microscope and record changes in appearance. Observations should be made at intervals not more than 24 hours apart, preferably every 12 hours.
>
> *Day 4.* Continue observations and recording of data.
>
> *Day 5.* Make final observations and begin to write paper.
>
> *Day 6.* Finish writing paper. Begin typing final draft. Ink in all drawings and graphs.
>
> *Day 7.* Finish paper.

It will be impossible for you to make up hypotheses without knowing *facts* about the phenomena you are studying. It will therefore be necessary for you to go to the library to do some preliminary reading before you are ready to make up your hypotheses. You might begin by reading the appropriate sections in the textbooks referred to at the end of this exercise. The most valid information (next to actual observation) will be found in biological journals, which report on scientific investigations and present original data.

In the **Introduction** section of your report, you will refer to the research reported in three books and/or scientific papers, stating how the facts uncovered in these publications led to your hypotheses. You should not need more than two or three hours of concerted effort reading the literature. (Most of the appropriate references will be on reserve in the library. Ask for each journal by its specific title, volume number, and date, as you see them on the chalkboard.)

After your introduction, you will state your hypotheses. The next section should be entitled **Discussion**. It will include a clear statement of your **experiments**, the **variables** involved, how your **control** affected these variables, and the **data** obtained. You may include charts, graphs, and drawings in this section. The final portion of your report will be entitled **Conclusions**. Here you will interpret your data and make whatever generalizations seem appropriate. Include a **Literature Cited** section. The report should be typed (double-spaced) and all drawings should be done in India ink.

There will be no formal laboratory sessions this week. The laboratory will be open all day for the performance of experiments and the recording of data. Your instructor will explain the methods of storage for your experimental setups. He will also tell you how to arrange for the use of the laboratory to do your work.

Some Useful Information

Brine shrimp can survive in seawater which has a salinity of 35 parts per thousand or 3.5%. Survival at this salinity does not mean that this is the *optimal* salinity; in fact, you have no evidence at this point that brine shrimp eggs will even hatch in seawater.

The way to prepare water of any salinity is to weigh out the appropriate number of grams of noniodized salt and place it into a 100 ml graduated cylinder. Add distilled water to the 100 ml mark. Thirty-five parts per thousand is the same as 3.5 parts per hundred (3.5%), so a seawater solution would be 3.5 grams of salt in 100 ml of water.

If you wish to test the influence of a pollutant, you probably will go to the library to look up the general range of pollution effects. You should document your reason for using a particular range in this manner: "Jones (1972) found that a concentration of ten parts per million of substance X caused distortions in developing embryos of organism Y. I therefore used solutions in a range of ten parts per hundred thousand to ten parts per billion to make sure I included the complete range of potential influences."

To make a solution of a particular concentration of pollutant, prepare a salt solution of the appropriate concentration as described above. (Find out the optimal salt concentration from your reading.) Place the appropriate number of ml of pollutant in a 100 ml graduated cylinder and add salt water to the 100 ml mark. This yields a parts per hundred solution.

Example: To prepare a ten parts per hundred solution, place 10 ml of pollutant in the cylinder and add *salt* water to the 100 ml mark. Yield: a ten parts per hundred solution (10 pph) of pollutant in salt water.

To prepare a ten parts per thousand solution (10 ppt), place 10 ml of the 10 pph solution in the cylinder and add salt water to the 100 ml level. Successive dilutions in this manner will yield parts per ten thousand (pptt), parts per hundred thousand (ppht), etc.

In order to test the effects of your experimental variables against the control, you will have to compare **hatching rates**. This means you must introduce some sort of quantitative measure. That is why it is so important for you to count the number of eggs you have in each dish under the dissecting microscope, and to regulate the number so that each dish has the same number of eggs. You can remove excess eggs with a nose dropper pipet. Each dish should have a large enough number of eggs to preclude errors due to small sample size.

Note that there is a great deal of creativity involved in this experiment. Think up a variable you really want to know about. In past years students have tested an amazing (and often blood-curdling) array of variables. Try to use an original variable that no one has ever tested before — or one which has relevance.

Caution: Observe your eggs at least once a day under the dissecting microscope. The hatched larvae are so small that they often cannot be seen with the naked eye.

BOOKS TO USE FOR GENERAL REFERENCE

Barnes, Robert, *Invertebrate Zoology*, 4th ed. Saunders, Philadelphia, 1981, pp. 675, 678.

Barrett, R. E., "Raise your own brine shrimp," in L. Pringle (ed.), *Discovering Nature Indoors.* Natural History Press, Garden City, NY, 1970, pp. 95-99.

Barth, Robert H., and Robert E. Broshears, *The Invertebrate World.* Saunders, Philadelphia, 1982, pp. 374, 382, 384.

Brown, Frank A. (ed.), *Selected Invertebrate Types.* Wiley, New York, 1950, p. 394.

Engeman, Joseph G., and Robert W. Hegner, *Invertebrate Zoology,* 3rd ed. Macmillan, New York, 1981, pp. 520, 704.

Galtsoff, Paul, et al., *Culture Methods for Invertebrate Animals.* Comstock, Ithaca, NY, 1937; republished by Dover Publications, New York, 1959, p. 205.

Hunt, J. D. (ed.), *Marine Organisms in Science Teaching.* Sea Grant College Program, Texas A&M University, 1980, 192 pp.

Institute of Laboratory Animal Resources, Assembly of Life Sciences, National Research Council, *Laboratory Animal Management, Marine Invertebrates.* National Academy Press, Washington, DC, 1981, 382 pp.

Meglitsch, Paul, *Invertebrate Zoology*, 2nd ed. Oxford University Press, New York, 1972, pp. 522, 524, 591.

Orlans, B. F., *Animal Care from Protozoa to Small Mammals.* Addison Wesley, Menlo Park, CA., 1977, pp. 39, 101-106, 118, 120-22, 172-73.

Schneider, E., and L. F. Whitney, *The Complete Guide to Tropical Fishes.* Thomas Nelson, New York, 1957, pp. 143, 149, 166-72, 189-93.

Welsh, J. H., and R. I. Smith, *Laboratory Exercises in Invertebrate Physiology.* Burgess, Minneapolis, 1947, pp. 42-43, 124.

29 Reducing a Complex Problem to Several Simpler Ones: The Subproblem

Population Dynamics in Open and Closed Systems

One of the most difficult tasks of the scientist is the reduction of his problem to a form which can be investigated. Suppose, for example, that a medical researcher is interested in finding a cure for cancer. No scientist will work on a problem like "What is the cure for cancer?" Instead, you might find one individual studying the induction of mammary tumors in mice, while another tries to find out whether or not a virus causes cancer in hamsters. These men are cancer researchers; they are studying aspects of the total problem expressed in terms amenable to investigation. They are asking questions which they feel they will be able to answer, rather than attacking a complex problem with so many variables that adequate control is impossible. Together, all the researchers will contribute to the total problem, the search for a cure for cancer.

It is often necessary to break a problem down into its component subproblems. The combined answers of these subproblems represent the solution of the major problem. In other words, a problem like "What kind of automobile should I buy?" cannot be solved if expressed in such a broad form. Instead, it must be stated so as to suggest activities which will result in a solution. Some appropriate subproblems might be

(1) How much money can I afford to spend on a car?
(2) What cars are available in my price range?
(3) Am I interested in a new car exclusively, or should I also consider a used car?
(4) What performance characteristics do I consider more important in a car? (For example, a great deal of power versus good gasoline mileage.)

Together, the answers to these subproblems represent the answer to the basic problem.

You will be able to see the virtues of breaking your problems down into several more easily handled subproblems in today's laboratory exercise.

PRELIMINARY INFORMATION

There are three general aspects of Darwin's theory of evolution—overproduction, competition for survival, and natural selection. **Natural selection** refers to the ability of the environment to "select" certain members of a species for survival, while others, unable to compete, do not live long enough to contribute their characteristics to the gene pool and thus become extinct. Species are constantly being "improved upon" or regulated as the changing environment culls out those individuals who are not maximally adapted to it. Today we will examine two populations of organisms in order to determine the interaction between the environment and those organisms.

Each group of students will receive 1000 cm^2 of turf (or 4000 cm^3, as the piece is 4 cm thick). Your group will also receive 1 cubic centimeter (1 cm^3 or 1 cc) of a culture of the oatmeal nematode, *Panagrellis* sp., or the vinegar nematode, *Turbatrix aceti*. The turf represents a naturally occurring animal and plant community. The nematode culture consists of water, oatmeal (or vinegar, in the case of the vinegar nematode), the nematodes, and whatever else can gain access to the culture from the air. The environment is "artificial" in the sense that we have created it in the laboratory.

■ THE PROBLEM: *Which contains a larger animal population, per unit of area, a natural turf community or a laboratory culture of nematodes? Does this relationship remain stable over time?*

□ Subproblem 1: *To determine the number of each kind of animal and the total number of animals per cubic centimeter of turf.*

At the front of the room is a piece of turf 1 m^2 in area by 4 cm thick. It is cut into sections 10 cm wide by 100 cm long. Bring a pan to the front table and obtain one section for your group. (Each group of four students will have one 10 X 100 X 4 cm section of turf.) At your table cut off a piece 2 cm long by 1 cm wide by 1 cm deep (2 cm^3). Include roots. Break up thoroughly with your hands and place in a 200 ml beaker. Remove all visible animals from this piece of turf and place them in a vial or 50 ml beaker of 50% alcohol. Spiders, centipedes, or millipedes can be collected by allowing them to climb up the side of a test tube and then shaking them into the vial or beaker. Put this container aside while you set up the Baermann apparatus.

Use of the Baermann Apparatus

Each team of four students will obtain a ring stand with ring attached, a funnel, a small dish or 50 ml beaker, a square of unbleached muslin cloth, a 4 cm piece of rubber tubing with clamp, a rubber band, and a dropping bottle containing 50% alcohol.

Add 150 ml of lukewarm water to the beaker containing the 2 cm^3 sample of turf and stir. Cover the top of the beaker with a single layer of cloth and attach securely with the rubber band. Do the rest of this procedure at the sink or in a tray or pan.

Prepare a funnel by adding a 4 cm piece of rubber tubing to the bottom and place the funnel in a ring stand as shown in Figure 29–1. Close the tubing with a pinch clamp and pour into the funnel enough lukewarm water that the water level will be at least 1 cm above the cloth when the beaker is inverted into the funnel. Gently invert the beaker and place it in the funnel. If necessary, add warm water until the level is about 1 cm above the level of the cloth.

Wait 30 minutes before drawing off enough water to just cover the bottom of a syracuse dish or 100 ml beaker. Examine the sample under a dissecting microscope. Systematically count the animals, starting from one side of the dish and progressing to the other. After recording your count below, wash out your dish and replace it under the Baermann apparatus. Remove and examine another sample every half hour for the next 1½ hours or more. Record your data below.

First half hour_____

Second half hour_____

Third half hour_____

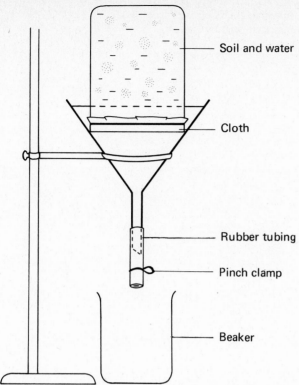

- Soil and water

- Cloth

- Rubber tubing

- Pinch clamp

- Beaker

FIGURE 29-1 Baermann apparatus.

Fourth half hour_____

Additional half hours_____

_____

Add counts to get total. Record here. _____

Place your total count on the chart on the chalkboard at the front of the room. Your instructor will divide the sum of the total counts by the number of setups in the room to

determine the average number of organisms per 2 cm³ sample. Write this number here_____ and in step (3) of the computation section below. Divide by 2 as instructed in the computation section to find the number of microscopic animals in 1 cm³ of turf.

While waiting to collect samples from the Baermann apparatus at half-hour intervals, perform the hand examination described below.

Examination of Soil by Hand

While the Baermann apparatus is collecting organisms, divide up the remainder of your piece of turf so that each member of your team has a chunk. Place each piece of turf on a large section of paper towel or in a pan. Each member of the team will break apart his/her chunk, searching carefully for animals living in the soil or in the roots of the plants. Add these animals to the vial or 50 ml beaker of 50% alcohol. Count the number of each kind of animal in the vial or beaker. Record your data in Table 29-1. If you are not sure of the name of an animal, consult pages 421–424.

**TABLE 29-1. Macroscopic Organisms from Turf
(Animals Removed by Hand)**

Specific Types of Organisms (include phylum name)	*Number of Organisms*

Example: Earthworm (Phylum Annelida)

Total [   ]

Computation of the Number of Animals per Cubic Centimeter of Turf

Compute the number of animals per cm^3 of turf as follows.

(1) Record the total number of animals from Table 29-1. _____
(2) Divide by total volume of turf.

$$\frac{\text{Number of animals} \rule{2cm}{0.4pt}}{4000 \text{ cm}^3} = \boxed{} \text{ (A)}$$

Number of macroscopic
animals per cm^3 of turf

(3) Divide class average of animals in Baermann apparatus by 2.

$$\frac{\text{Number of animals} \rule{2cm}{0.4pt}}{2} = \boxed{} \text{ (B)}$$

Number of microscopic
animals per cm^3 of turf

(4) Add numbers of macroscopic and microscopic animals.

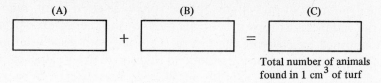

Total number of animals
found in 1 cm³ of turf

Give your answer for the total number of animals per cubic centimeter of turf (C) to your instructor, who will compute the class average on the chalkboard. Class average _____ (Use this number for any further computations.)

1. Is there anything wrong with the sampling technique? _____

Explain or comment. _____

_____

2. How close do you believe our estimate of the population per cubic centimeter is to the actual number (check one)? very close _____ close _____ not close _____

Why? _____

_____

□ **Subproblem 2:** *To determine the number of nematodes per cubic centimeter of nematode culture.*

Using a transfer pipet, place exactly 1 drop of the nematode culture (to be found at your table) on a glass slide. Spread the drop out until it covers about 2 cm² of the slide. If the medium is too opaque for you to see the worms clearly, add 1 or 2 drops of water. Count the number of worms under a dissecting microscope. Examine the square of medium on the slide in a systematic manner, starting from the upper left-hand corner. Divide the square into three columns, counting the worm population in each third by moving your eye first down the left-hand edge of the square, then up the middle, then down the right-hand edge as shown in Figure 29-2.

Record the number of nematodes per drop of culture. _____

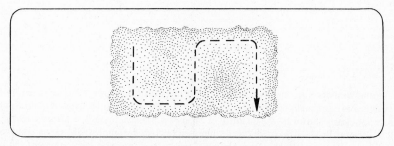

FIGURE 29-2 Pattern for examining slide.

To calculate the number of drops per cubic centimeter of culture, fill a 1 ml transfer pipet with water to the 1 ml mark (1 ml = 1 cm³ ; liquid volume is given in ml, solid volume in cm³) and then count the number of drops it contains by allowing the water to drip slowly from its end. This is accomplished by covering the top of the pipet with your forefinger and releasing it gently until the water flows out of the other end in discrete drops. If all four students of the group perform this operation and compute an average, the result will be a better estimate.

Record the average number of drops in 1 cm³ of culture. _____

Multiply the number of nematodes per drop times the number of drops per cubic centi-

meter to get the number of nematodes per cubic centimeter._____

Results

Compare the number of animals per cubic centimeter of nematode culture _____

with the number of animals per cubic centimeter of turf _____ .*

[Sometimes, if there is not enough class time devoted to allowing the Baermann apparatus to operate, the number of microscopic animals in the soil is artificially small. At the front table is a Baermann apparatus which was set up yesterday and allowed to run overnight. Your instructor may wish you to use a corrected number which includes this long-run information; if so, you will find the number of animals obtained overnight recorded on a card in front of the apparatus. If this is the case, correct your total count of soil animals (C) by adding your number of macroscopic animals (A) to the overnight count of the Baermann apparatus. Total

number _____ (new C)]

Whether you use your original data or the new total, you should be able to make a generalization concerning the number of animals in a natural population (turf) versus a laboratory culture (oatmeal or vinegar nematodes).

3. What is your generalization? Is there a difference in the density of the two populations?

_____

4. What possible reasons can account for this?

*Our computations to determine the productivity of portions of the environment are based on the number of organisms present. This measure is not usually used by ecologists because the amount of animal or plant *tissue* supported by a part of the environment is not clearly shown by the quantity of organisms present. Instead, ecologists use the *biomass* or weight of the tissue, taking into account the fact that five beetles represent a great deal more living tissue than would fifty microscopic nematodes. We use the number of organisms rather than biomass because we lack the equipment and time to determine biomass and because, in this instance, the principles governing both measures are in fact the same.

Apparently there are more animals per cubic centimeter of culture medium than there are in the natural system comprising the turf community. This is not surprising since we know that more or less artificial culturing methods in agriculture, such as growing vegetables in special liquid baths (hydroponics) or raising chickens in boxes on highly mechanized chicken farms, produce huge populations.

The problem we now face is to determine whether or not the relationship we have found between turf and nematode culture populations can continue through time. There are many factors which affect population size. These can include availability of water and food, fecundity (ability to produce offspring) of each species, temperature, predation, disease, and accumulation of wastes which can reach toxic levels. In an ecological situation, only one of these factors will actually limit the size of a population. This is called the limiting factor. While any of the variables mentioned *might* limit the size of a population, only one will reach critical proportions *first*—and that one will be the limiting factor. In one particular habitat or ecosystem, the limiting factor might be different from another.

5. For example, what is the probable limiting factor in a desert? _____

In the Arctic? _____

□ **Subproblem 3:** *To determine the possible limiting factors which affect the stability and size of the turf and nematode culture populations.*

The nematode culture contains water, oatmeal (or vinegar), and nematodes. Unlimited amounts of air can enter through the lid. It has been shown that these nematodes do not eat oatmeal (or vinegar).

6. If this is the case, how can they survive? (It might be worthwhile for you to examine

the culture carefully, even to smell it.) State your hypothesis. _____

_____

7. The turf community contains many different species of animals: ants, earthworms, spiders, nematodes, beetles, etc. List possible food sources for these animals.

_____

_____

_____

_____

_____

_____

_____

8. Compare the smell of the nematode colony to the smell of the turf. _____

_____

_____

Below are lists of possible limiting factors in each community. Using the number 1 for most important, and successive numbers for less and less important, indicate the likelihood of each factor being *the* limiting factor in each community.

Turf Community	*Nematode Culture*
Predation	Predation
Disease	Disease
Food	Food
Water	Water
Temperature	Temperature
Fecundity	Fecundity
Waste accumulation	Waste accumulation

9. Explain the relative importance of each factor below.

Suppose that additional samplings are made at weekly intervals as follows.

(1) A fresh sample of turf is dug up and the animals counted.
(2) A sample of nematode culture is counted. *The oatmeal or vinegar has not been replenished.*

Decide whether more, fewer, or the same number (approximately) will be found in each population in the future.

10. Number of animals in turf (check one): more ____ fewer ____ same ____

11. Number of animals in nematode culture: more ____ fewer ____ same ____

12. Explain why you chose the particular alternative you did in each case. What are the facts, thoughts, and reasons which permitted you to make your hypotheses?

In the graphs below draw a line representing each of your hypotheses concerning the fluctuations in the turf and culture populations.

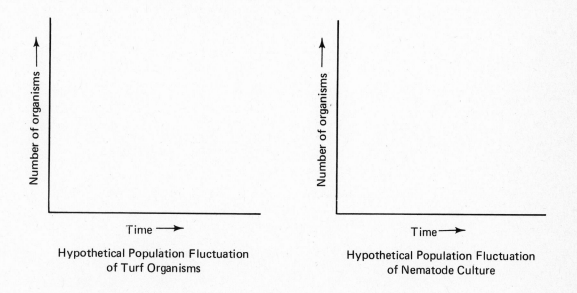

Hypothetical Population Fluctuation
of Turf Organisms

Hypothetical Population Fluctuation
of Nematode Culture

In the space below and on the next page, write the problem you have been trying to solve, the subproblems, and the solutions to the problem and subproblems.

Make sure that you clean up all loose soil as directed by your instructor.

30 Understanding Contemporary Biological Problems I

Human Ecology 1: Effects of Overpopulation

The term "population explosion" is one which we have learned to live with. From the urgency in the voices of those few biologists who have reached the mass media we realize that the problem is of major importance. This exercise will treat the problem purely in biological terms—with the understanding that the reality of the threat is much more complex. Each of the political, social, and economic aspects of the problem of overpopulation is literally overwhelming.

Consider the following facts:

Fact: In 1954 United Nations demographers predicted that by 1980 the Earth's population would be about 3.6 billion people. In 1965 the projection had to be changed because the world population was already approaching 3.6 billion. The actual number reached in 1980 was about 4.3 billion. Somehow the rate of human population growth had risen about 20% above its horrifyingly high predicted rate.

Fact: The growth rate of the human population is 2.1% today and is rising fast. If it gets to 3%, the world population will double every 23 years. At the present rate there will be 7.5 billion people on the Earth by the year 2000; within your lifetime that number may again double to 15 billion. The most optimistic projections place the limits on the Earth's ability to sustain its human population at around 10 billion.

Fact: The United States and certain European countries have virtually reached zero population growth (ZPG). Most population increases are found in the overcrowded, impoverished nations of the Third World.

Fact: Aid for the poorer nations from the richer has decreased from 0.8% to 0.6% of the incomes of the rich nations in the past few years. The "have" nations appear to be getting tired of helping the "have-nots."

Fact: Desalination (removing salt from ocean water to make fresh water) will not become economically feasible for farming within the foreseeable future. This means we can forget about "making deserts bloom" to feed expanding human populations.

Fact: The "green revolution," whereby advanced nations have shown the underdeveloped countries how to grow crops with many times the traditional yield, has resulted in an almost miraculous increase in productivity in many impoverished nations. Their increased production has, within one generation, been more than compensated for by an even greater increase in consumption, due to the increased survival of infants and greater longevity of adults. More people = less food. Thus, on a per capita basis, many of these nations are poorer now than before the green revolution.

PRELIMINARY INFORMATION

A single pair of Atlantic codfish and their descendants, reproducing without hindrance, would in six years fill the Atlantic Ocean with their packed bodies.

The equation describing their rate of increase would be

$$\text{Rate of increase in number of individuals} = \left(\text{Average birth rate} - \text{Average death rate}\right) \times \text{Number of individuals at a given time}$$

If you want to calculate the rate of increase of a colony of 1000 mice, with an average birth rate of 30 young per 100 mice per year, or 30%, and an average death rate of 10 per hundred per year, or 10%, the rate of increase (really the excess of births over deaths) would be (.30 – .10) × 1000 = .20 × 1000 = 200, or an increase of 200 mice after one year.

But there are other factors which affect population size. Together, these are called "the environment." They are divided into two kinds of influences, those that are **density dependent** and those that are **density independent**. Density-dependent factors include all those aspects of the environment influencing population size which increase or decrease in effect according to the number of organisms per unit of area; for example, if conditions become crowded, diseases will be more readily transmitted. Communicable disease is usually a density-dependent factor; so is the availability of food.

Density-independent factors are those which would have the same effect no matter what the density of the population. Weather is an example of a density-independent factor. If there is no rainy season on the African plains, populations will decline whether or not they are crowded. However, it is difficult to separate the two kinds of factors because under most circumstances even a drought brings *some* rain. With a small amount of water, the animals would be forced to compete for it, so that low population densities would be an advantage. Under any circumstances, there is a limit to the amount of food, water, etc., available in any habitat. Thus each habitat has a **carrying capacity**—the maximum number of organisms it can support.

Our equation representing the rate of increase in a population must now be changed to include the environment.

$$\text{Rate of increase in number of individuals} = \left(\text{Average birth rate} - \text{Average death rate}\right) \times \left(\frac{\text{Carrying capacity of environment} - \text{Number of individuals}}{\text{Carrying capacity of environment}}\right) \times \text{Number of individuals at a given time}$$

Suppose that we analyze our mouse colony, introducing the environmental aspect. Let us follow the hypothetical growth of the colony over a period of time. Initially there is a low population (10 mice) in comparison with the carrying capacity of the environment. Succeeding months show an increase in population size. There are three examples where the number of mice is nearly half of the carrying capacity, and three examples where the population almost equals the carrying capacity. Using the following population sizes (number of individuals), solve the equation for the rate of increase.

Month	Population Size
0	Very low population (10 mice)
	Very low population (50 mice)
	Low population (100 mice)
	Medium low population (300 mice)
	Medium population (500 mice)
	Medium high population (700 mice)
	High population (900 mice)
	Very high population (950 mice)
36	Very high population (990 mice)

You also need the following data to solve the equation:

Average birth rate: 30% or 300 per thousand mice per year
Average death rate: 10% or 100 per thousand mice per year
Carrying capacity of the environment: 1000 mice

Example:

$$
\begin{array}{l}
\text{Rate of increase in number of individuals} = \left(\text{Average birth rate} - \text{Average death rate} \right) \times \left(\dfrac{\text{Carrying capacity of environment} - \text{Number of individuals}}{\text{Carrying capacity of environment}} \right) \times \text{Number of individuals}
\end{array}
$$

$$
= .30 - .10 \times \left(\frac{1000 - 10}{1000} \right) \times 10
$$

$$
= .20 \times \left(\frac{990}{1000} \quad \text{or} \quad .99 \right) \times 10
$$

$$
= .20 \times .99 \times 10
$$

$$
= 1.98
$$

Answer: 1.98 mice increase per unit of time for a population of ten mice.

Calculate the rate of increase for the remaining eight population sizes.

Plot the rates of increase for different population sizes on the graph on the opposite page. Turn the book sideways so that the horizontal axis is the long axis. Label the horizontal axis (abscissa) "Number of mice." Each small box represents 10 mice.

Label the vertical axis (ordinate) "Rate of increase." Each small box represents 1 mouse.

1. What should be the title of the graph?_____

_____

Write the title across the top and plot the graph now.

2. Examine your graph. When is the rate of increase most rapid? _____

When is the rate of increase least rapid? _____

3. The graph will show *two* cases where the rate of increase is least rapid. Explain the reason for the slow increase in population in each case.

a. First case._____

_____

_____

_____

b. Second case._____

_____

_____

_____

These two answers are critical to an understanding of the entire exercise. If you are not sure of them, consult your neighbor and/or the instructor!

In the preceding example you calculated the **rate of increase** in a population of mice. You found that the rate of increase declined as the size of the population approached the carrying capacity of the environment. Now let us consider the actual **population size** of a colony of mice placed on a small island near the shores of a lake. The island contains no mice but is regularly visited by wildcats and foxes.

Use the data on page 382 to plot your curve on a second graph (page 383). Turn the book sideways so that the abscissa (horizontal axis) is the long axis. It should be labeled "Months from date of colonization."

The ordinate (vertical axis) should be labeled "Number of mice." Each small box should represent 10 mice, from 0 to 700. Note that this is the **total number** of mice, not the rate of increase.) Mark off intervals of 3 months every 10 boxes, so that the left-hand corner of the graph is marked 0 (date of colonization) and the number 3 (for 3 months) is located 10 small boxes from the zero on the horizontal axis.

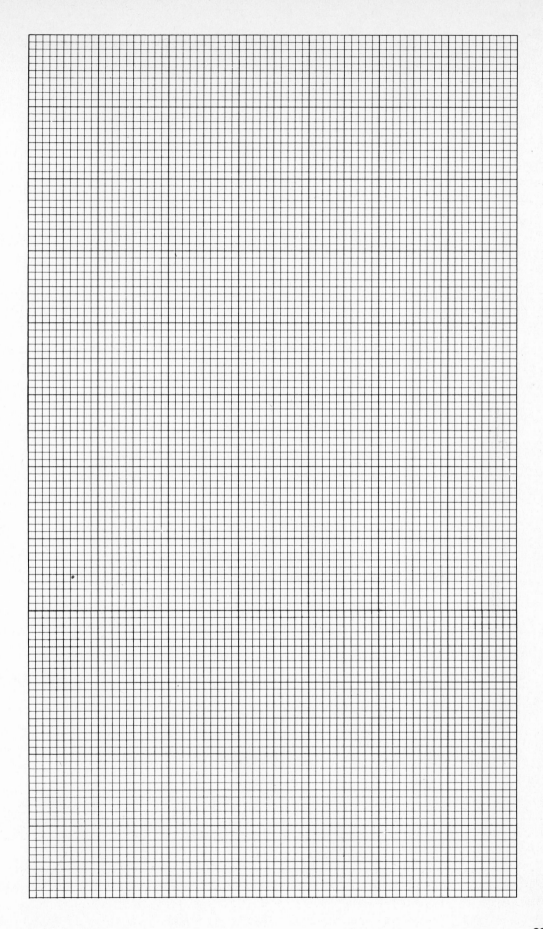

Months after Colonization	Number of Mice
0	10
3	32
6	86
9	247
12	546
15	601
18	583
21	590
24	520
27	562
30	608
33	618
36	570

Plot your curve now.

4. What is the title of this graph?_____

_____

Write the title across the top of the graph.

5. Describe what has happened to the population of mice on the island.

6. Why did this happen? Use the following terms in your explanation: carrying capacity, rate of increase, density-dependent factors.

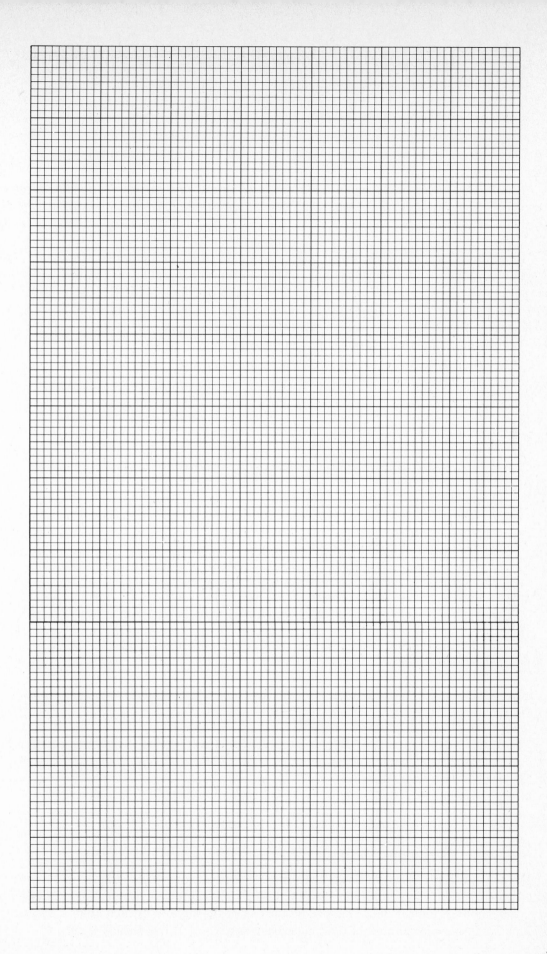

This laboratory exercise is concerned with the effect on the size of the population of an increase in the carrying capacity of the environment. You will be given two micro-habitats or "miniature worlds." They will consist of vials, one of which will be three times the size of the other and will contain three times the amount of food. Your problem is to study the effects of tripling the size and food supply of the "world" in which you will place six fruit flies.

■ THE PROBLEM: *What will happen to the population size of a colony of fruit flies given a habitat three times the size and with three times the food of the control colony?*

Write in the space below what you believe will happen in both colonies if they are left alone without adding or removing anything. What do you think will happen in the large vial? In the small vial? Compare the two. Indicate what the population curve in each vial might look like; and mention, in general terms, the relationship between the number of flies in the large vial versus those in the small vial. Draw your hypothesized population curves for the large and small vials. Superimpose one curve on the other.

□ Hypothesis:

7. How many times more fruit flies do you think will be in the large vial than in the small

vial at the peak of each population? ————————————————————————————

Procedure

Work in pairs. Prepare the vials as follows.

(1)   Fill the cap of the small vial with dry *Drosophila* food (0.84 gram). Level with a straightedge. Pour into the small vial. Add 1½ capfuls of water. Allow to set for 5 minutes. Repeat the process using the cap of the large vial, which holds 2.52 grams, three times the amount in the smaller vial. Again add 1½ capfuls of water.

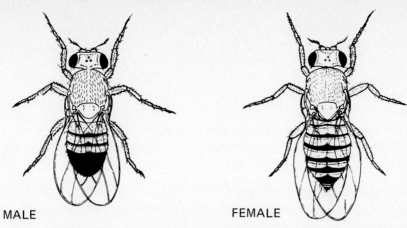

FIGURE 30–1 How to differentiate between male and female fruit flies. Note that abdomen of male is more rounded than that of female and has a black posterior third. Females are larger than males and have five alternating light and dark bands on the abdomen.

    (2)   Add a few grains of yeast to the small vial. To keep quantities constant, add approximately three times the number of yeast grains to the large vial. This does not have to be exact.

    (3)   Allow to set 5 minutes.

While allowing the food to set, obtain some anesthetized flies from your instructor. Pour the flies on a white 3 × 5 card, and place the card on the stage of a binocular microscope. Using a camel's-hair brush, separate the flies so that you can determine their sex (see Figure 30–1). If the flies begin to recover too rapidly, cover them for a minute with the lid of a petri dish to which is affixed a gauze pad with a drop of ether.

Place the vials on their sides and slide three male and three female flies into each vial. Cover. Allow the vials to remain on their sides until the flies have recovered and are crawling vigorously. Gently right the vials making sure the flies do not fall into the food and become mired in it.

Take the vials home or to your dormitory room and leave them in a secure place, out of direct sunlight. It will not be necessary to open the vials at any time during the entire investigation. Air will enter through the tiny holes in the lid. Record the number of *adult* flies present every three days. Figure 30–2 depicts the four stages in the life cycle of a fly.

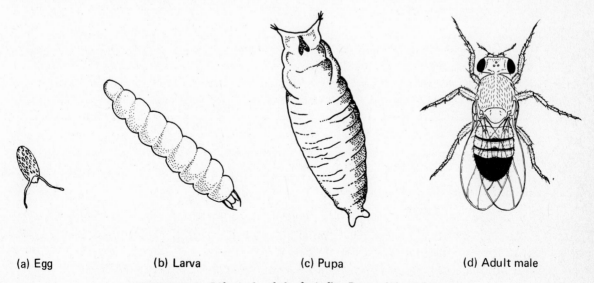

(a) Egg   (b) Larva   (c) Pupa   (d) Adult male

FIGURE 30–2 Life cycle of the fruit fly, *Drosophila melanogaster.*

After a while, so many flies will be in the vials that you will have difficulty counting them. Placing the vials in a refrigerator for an hour will cause the flies to become sluggish, making counting easy. Lay the vials on their sides, so that when the flies relax, they will not fall into the medium. If access to a refrigerator is difficult, bring your vials to the laboratory twice a week. Your instructor will place them in a refrigerator for you.

If you have difficulty seeing the flies because the sides of the vial become dirty, hold the vial against the light from a window and the flies will become visible.

When your investigation is completed, or on the date specified by your instructor, turn in a short report, with graphs. Make sure to include the following information.

(1) An *introduction* with references to other literature (in this case, textbooks will be acceptable) you have read relating to the problem of overpopulation.

(2) The *problem*.

(3) Your original *hypothesis*.

(4) Your *data* in both tabular and graph form. Make sure both curves (for small and large vials) are on the same graph.

(5) The *interpretation of your graph and numerical data*. Compare your curves to the graph of the mouse population on the island. Explain any difference.

(6) Your *conclusions*.

The report should be in journal style. Your instructor will have available research papers to use as models.

31 Understanding Contemporary Biological Problems II

Human Ecology 2: Human Activities and Water Pollution

Inherent in the American free enterprise system is the tendency of governmental agencies to exert minimal direction on the industrial sector. This tradition has served America well; it has led to unprecedented industrial growth and prosperity. But the mainstay of this prosperity has been the availability of inexpensive sources of energy (an abundance of coal, oil, and water power) and the presence of large areas of land and water upon and into which incredible quantities of industrial waste have been thrown. These almost inexhaustible resources are now depleted. America's total oil reserves cannot supply our energy requirements for more than five or ten years; coal is not suitable for many uses; the supply of available rivers to be dammed for hydroelectric power has diminished since the ecological impact of the impounding of waters has become better known.

The pollution from industrial, farm, and domestic by-products and wastes has led to the destruction of such large segments of the environment that normal and healthy living conditions for man as well as for many animals and plants are threatened.

Pollution can be defined as the introduction of substances produced by human activity into the environment in such large quantities as to interfere with normal processes and equilibria. Pollutants can be wastes, such as sulfuric acid or mercury, from industrial processes, or they can be excessive amounts of fertilizer or phosphates from detergents washed into lakes and causing unnatural "blooms" of algae. Pollution, it must be emphasized, is simply the presence of substances in such abundance as to render inadequate the natural cycles which have evolved over millions of years. Nitrates and phosphates are the "life-blood" of a lake. Every lake depends on yearly inundations of these minerals which are leached out of the soil by spring rains and brought to the surface by the turnover of the lake water every spring.

The introduction of nitrates and phosphates stimulates the growth of algae in the lake, just as a spring application of fertilizer stimulates the growth of grass on a lawn or crops in a farmer's field. However, if some of the farmer's fertilizer washes off his land and enters feeder streams, and if it is further augmented by phosphates from countless sinkfuls of dishwater, the amount of fertilizer will be excessive. It will cause unnaturally rapid growth of algae in the lake waters which will, in turn, overtax the natural regulators of plant growth, the herbivorous animals (especially the zooplankton). The normal population of consumers will be unable to keep the number of plants down. Huge numbers of plant cells will live through their life cycles and die, sinking to the bottom where they oxidize. This process of disintegration uses up much of the available oxygen, causing death by anoxia (oxygen deprivation) of the animals in the lake and reducing still further the ability of the lake to cope with the unnatural growth of its own plants. Eventually the lake "dies." Its capacity to maintain itself is shut off. Its fish and zooplankton will die or be drastically diminished in numbers. The lake will stay in this condition permanently, or until the original level of plant production is re-established.

Recovery might never happen if the watershed of the lake is occupied by cities (sources of high phosphate detergents and domestic sewage) and farms (sources of fertilizer run-off). Such is the ever-present danger for Lake Erie, the twelfth largest lake in the world, on whose shores lie the cities of Buffalo, Cleveland, and Toledo.

The problem appears to lie in the *concentrations* of substances introduced into the environment by man. There are natural levels of phosphates, nitrates, mercury, salts, hydrocarbons, etc., in surface waters. When the concentrations are increased beyond the capacity of the system to neutralize their influence, an unnatural and potentially dangerous condition develops.

The purpose of this laboratory exercise is to study the effects of different concentrations of substances on aquatic organisms in order to find out when these concentrations reach a point of becoming pollutants.

PRELIMINARY INFORMATION

A problem exists in determining the toxicity of substances placed in water. How will we know if something is harmful? What standards exist? If we pour a chemical into a lake and everything dies, there is no question about its toxicity. If only a few organisms die, how do we know that their death stems from the action of the chemical? Furthermore, all poisons have varying effects on a population. For every organism that dies, a considerable number suffer discomfort in varying degrees.

The generally accepted criterion for toxicity is called the LD_{50}: the single-dose quantity of a substance that will kill 50% of the test population in a given amount of time.

Moran et al. describe the LD_{50} as follows.

It is usually given in the units of milligrams of poison per kilogram of body weight. For example, the chemical that causes food poisoning (botulism) has an LD_{50} of 0.0014. Hence if we feed the poison to ten men each weighing 220 pounds (100 kilograms), a dose of 0.000049 ounce (0.0014 mg) would kill five of them.[*]

The concentrations for the LD_{50} vary with type of organism, mode of entry, age, sex, and many other factors, even climate. This problem is magnified by the fact that new chemicals and drugs are tried out on experimental animals such as mice and guinea pigs. The LD_{50} for mice in a temperature-controlled laboratory is worlds apart from that of an elderly woman in the tropics or a fetus in its mother's uterus. That is why the introduction of new drugs is so time-consuming. It is easy to understand why there is often a lack of agreement among experts as to what are the safe levels of drugs or chemicals that are to be introduced into the environment.

A final confusing note particularly applicable to insecticides is the fact that LD_{50}'s are computed on the basis of specific concentrations. If a farmer buys a bottle of insecticide with instructions to pour one pint into forty gallons of water to make the proper dilution, and he uses one pint in twenty gallons "to really get rid of the bugs," the original dilution which produced the LD_{50} is no longer applicable.

The difficulty of determining levels of toxicity can be understood better when one considers a huge factory on the bank of a river. Its wastes may change the pH of the river from 7.1 to 6.8. They may raise the river's temperature ½°C. They may introduce a few parts per million of mercury or sulfur into the water. It might be financially impossible or tremendously costly to reduce the levels of any of these substances, and the resulting increase in the cost of

*[*]J. M. Moran, M. D. Morgan, and J. H. Wiersma, *An Introduction to Environmental Sciences,* Little, Brown & Co., Boston, 1973, p. 25.

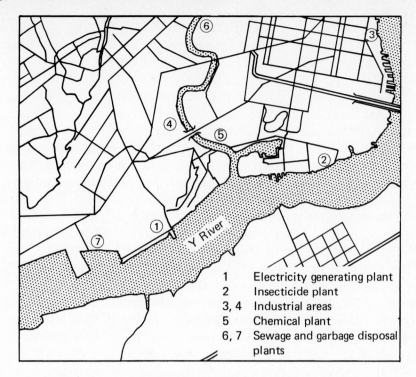

1   Electricity generating plant
2   Insecticide plant
3, 4 Industrial areas
5   Chemical plant
6, 7 Sewage and garbage disposal
      plants

FIGURE 31-1   Map of city X showing sources of pollution.

production usually goes into the retail price paid by the consumer. It is apparent that the interaction of the pollutants with one another and with the river must be determined, as well as the precise toxicity of each. If raising the pH of the river from 6.8 to 7.0 would cost billions of dollars over the years, the precise determination of the toxicity of the lower pH must be clearly demonstrable if pressures from industrial and consumer lobbyists to keep prices down are to be counteracted.

The problem you will deal with today is a real one. In the summer of 1967 a city, hereafter called X, faced an ecological crisis. The level of pollutants rose to a point which threatened life in the Y river. The sources of pollution and their relationship to the river are shown on the map (Figure 31-1)—and the levels at which the pollutants were found are given in Table 31-1. The crisis was described in an article which suggested that a barrier to the passage of fish and other organisms might be created as the river passed the city and received its pollutants. Heat from the cooling systems of electricity-generating plants combined with the naturally high temperatures of the river in July and August represented **thermal pollution.** Incompletely processed human wastes from inefficient sewage-disposal plants plus other organic wastes from slaughterhouses, etc., poured into the river and settled on the bottom, were oxidized, and used up what little oxygen was available in the warm water (generally, the

TABLE 31-1   Water Characteristics, Y River, July 27, 1967

Temperature (max.)	31°C (88° F)
Dissolved oxygen (bottom)	1 ppm
Ammonia nitrogen	15 ppm
Ortho phosphate	4.8 ppm
pH	6.8
Suspended organic matter	12.3 ppm
Insecticide residues	6 ppm

warmer the water, the more dissolved oxygen is driven out of it by the rapid movement of the water molecules). The capacity of organic materials to use up dissolved oxygen in the water is called biochemical oxygen demand (BOD). It is an important index of pollution. Substances which create a BOD are called **organic pollutants**. Finally, a number of industrial plants released a variety of chemicals into the river including insecticide by-products; sodium salts; nickel, cadmium, lead, and other metal ions used as catalysts and for other purposes; and sulfuric acid, a waste product of many industrial processes. These are **chemical pollutants**.

The particular problem was two-fold. Shad and striped bass, two important fish used for food, had migrated upriver in the spring on their annual spawning runs. The fish had little difficulty in passing the area around city X because they were large, healthy, and in their spawning prime and because the temperatures in the river were close to their coldest, originating from melting snow and ice which fed the river at its headwaters. But the eggs had hatched upstream, and by July the fish larvae had grown to fingerling length (2–5 cm) and were ready to migrate downstream to the ocean. In their path lay the city X area. Would the young fish be able to pass the barrier of pollutants? Would they succumb to the lack of oxygen and suffocate? Would they be able to stand the shock of the lowered pH (from the acid wastes) and the high temperature (from thermal pollution superimposed on the high summer temperature of the water)? Furthermore, would the natural food of the migrating fish, crustaceans and worms, be available in the long stretch of river adjacent to the city? Or would these tiny food animals, permanent residents of the polluted area, have succumbed, thus adding the burden of starvation to the already endangered army of fingerlings as they reached the city X area? This was the problem facing concerned city officials as temperatures mounted in a July heat wave.

■ THE PROBLEM: *What are the levels of thermal and chemical pollution (sulfuric acid and insecticide) which are harmful to two representative aquatic food organisms,* Daphnia *and* Tubifex?

Work in groups of four, each pair of students doing one of the two parts which follow. One pair will test the effect of insecticide on *Daphnia* and *Tubifex* (Subproblem 1). The other pair will test the effect of lowered pH (Subproblem 2). After both teams have set up their experiments, they will work together on Subproblem 3, testing the combined effects of insecticides, lowered pH, and thermal pollution.

Before beginning the experiments, all students will read through the procedures for Subproblems 1 and 2 in order to learn the necessary techniques.

□ Subproblem 1:   *What are the effects of various concentrations of insecticide on* Daphnia *and* Tubifex?

In the space following "Protocol" below and on the next page, write your experimental design, including the assignments for each member of the group. (Do this after you have read the procedures for Subproblems 1 and 2.)

Protocol:

Preparing Dilutions

To prepare dilutions of insecticide follow this procedure.

(1) Read the label on the bottle of insecticide to find out the concentration of its active ingredients, by volume.

(2) Your instructor will have available a bottle of insecticide diluted by placing 10 ml of the commercial preparation into 90 ml of aged water. Pour 10 ml of this preparation into your graduated cylinder. If the original concentration of insecticide was 12% active ingredients by volume (88% inert ingredients), that is the same as saying that there were 12 parts of active insecticide per 100 parts of solution, or 12 pph (parts per hundred). Your instructor has diluted the 12 pph solution by one tenth. What is

the dilution in this bottle? _____

(3) Add 90 ml of aged water to the 10 ml of insecticide in your graduated cylinder.

Place this in a bottle and label. What is the dilution of this bottle? _____

(4) Repeat your one tenth dilutions so that you have bottles containing the following dilutions.

(a) parts per thousand (use only with *Tubifex,* not with *Daphnia*) ($\times 10^3$)
(b) parts per ten thousand (use only with *Tubifex,* not with *Daphnia*) ($\times 10^4$)
(c) parts per hundred thousand ($\times 10^5$)
(d) parts per million ($\times 10^6$)
(e) parts per ten million ($\times 10^7$)
(f) parts per hundred million ($\times 10^8$)
(g) parts per billion ($\times 10^9$)
(h) parts per ten billion ($\times 10^{10}$)

Stop to reflect for a moment how small a quantity of insecticide there would be in water of a concentration of *12 parts per ten billion*!

From each bottle of diluted insecticide (beginning with (c), ppht, for *Daphnia)* pour enough solution so that the test tube is filled to within 1 cm of the top. Label each tube with the concentration (ppm, ppht, and so on) and draw two lines with a china-marking pencil to divide it into three zones—upper, middle, and lower—as shown in Figure 31-2. Add 10 *Daphnia* to each tube. This setup will allow you to test the effects of various concentrations of insecticide on the *Daphnia* in two ways. One way is to determine the LD_{50}.

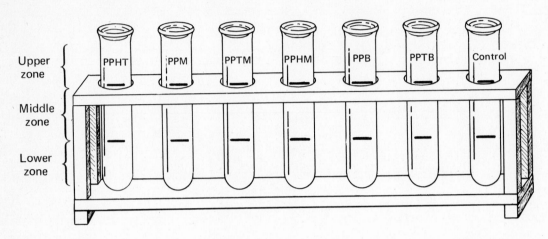

FIGURE 31-2   Setup for testing *Daphnia* for susceptibility to dilutions of insecticide.

Record the time when you have completed adding the *Daphnia* to the solutions in all test

tubes. Time: _____ After 1 hour has elapsed count the number of live *Daphnia* in each tube and make the proper entries in Table 31-2. Normally LD_{50}s are determined after 24 hours, 1 week, or even 1 month.

1. Explain why the LD_{50} may change with time. _____

_____

_____

_____

The other method of determining the effects of the insecticide is usually less quantitative, but may sometimes be more sensitive. Observe the behavior of the *Daphnia* to see if the insecticide is affecting them in a sublethal manner. Look for any change in behavior, such as preferred depth, rate of swimming, etc.

2. If the pollutant is not lethal, yet affects behavior, it may still lead to the disappearance

of the organisms in the affected area. Explain this. _____

_____

_____

TABLE 31-2   Data for Insecticide Experiments

A. Numbers of Dead Animals After 1 Hour

Dilution	Dead Daphnia	Dead Tubifex
ppt	—	
pptt	—	
ppht		
ppm		
pptm		
pphm		
ppb		
pptb		
Control		

B. Numbers of *Daphnia* at Each Level

Dilution	Level	3 min.	6 min.	9 min.	12 min.	15 min.	1 hour
ppht	Upper						
	Middle						
	Lower						
ppm	Upper						
	Middle						
	Lower						
pptm	Upper						
	Middle						
	Lower						
pphm	Upper						
	Middle						
	Lower						
ppb	Upper						
	Middle						
	Lower						
pptb	Upper						
	Middle						
	Lower						
Control	Upper						
	Middle						
	Lower						

One way of quantifying the observations would be to count the number of *Daphnia* in each of the three zones in your test tubes and compare this to the control. Three minutes after all *Daphnia* have been added to the tubes, record the number of *Daphnia* in each zone (upper, middle, lower) of each test tube in the appropriate column of Table 31-2. Continue recording these data after 6, 9, 12, 15, and 60 minutes have elapsed.

3. Discuss the pros and cons of this approach versus the LD_{50} approach. _____

_____

_____

_____

_____

4. Record here which dilution reached the LD_{50}. _____ Time exposed *_____

5. Record here which dilution produced clear-cut behavioral differences _____

Time exposed: _____

While waiting for results, set up small vials or petri dishes to test the *Tubifex*, as follows.
Add 20 ml of each dilution to a small petri dish or vial, beginning with parts per thousand (ppt). At your table you will find a finger bowl containing hundreds of worms in a mass. With a forceps remove a small clump of worms. If the worms are separated you can suck them into a nose-dropper pipet. Pick out and place in each dish approximately ten worms. It will be difficult to count the worms accurately; a good estimate will be acceptable. Determine the LD_{50}. It is too difficult to determine behavioral differences in worms, so the LD_{50} data will have to suffice.

6. Record here which dilution reached the LD_{50}._____ Time exposed _____

□ **Subproblem 2**   *What are the effects of a lowered pH on the survival of* Daphnia *and* Tubifex?

Pair two will set up five test tubes or vials and five small petri dishes or vials, plus controls, in the manner described for subproblem 1.
Use the 5% sulfuric acid to prepare solutions according to the technique described below. *Be very careful in the use of the acid.* Inform your instructor if acid spills on your clothes or skin.
Fill a nose-dropper pipet with 5% sulfuric acid. Place the tip of the pipet above the surface of the water in a test tube for *Daphnia* (see Figure 31–3) or in a petri dish for *Tubifex*. Mix with a glass stirring rod.
The amount of acidity is not usually expressed in parts per thousand or million. Instead the concentration of hydrogen ions is measured on an index of acidity. This index, called pH, is graduated from 0 to 14. Zero is most acidic, and 14 is most basic; pH 7 is neutral. Test the pH in your control tube or dish to determine its pH, using the litmus paper or pH meter provided. Your instructor will explain its use. Lower the pH of your experimental tubes and dishes by adding drops of acid until the pH is reduced by intervals of 0.4 pH unit (for example, if original water pH is 7.2, your solutions will be 6.8, 6.4, 6.0, 5.6, and 5.2). Once you have determined that five drops, for example, reduced the pH from 7.2 to 6.8, then ten drops

---

*This will probably be 1 hour unless your instructor has provided more time.

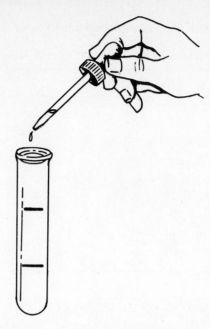

FIGURE 31-3 Adding dilute sulfuric acid to test tube.

would reduce the pH from 7.2 to 6.4, etc. Check with your litmus paper or pH meter. There should be five solutions plus the control.

Place ten *Daphnia* in each of the six test tubes and approximately ten *Tubifex* in each of the small petri dishes. Make your observations as described on pages 392 and 394 and fill in Table 31–3 (page 396). Wait 1 hour (preferably 4, 8, or even 24 hours) and determine the LD_{50}.

Daphnia: LD_{50} _____ Time exposed _____

Tubifex: LD_{50} _____ Time exposed _____

7. Explain why the LD_{50} may change with time. _____

_____

_____

_____

8. Examine the data for insecticide obtained by pair one and your data above. Are both

Daphnia and *Tubifex* equally susceptible to low pH and insecticide? _____

_____

_____

Make sure to copy the data from the other team so that both Table 31–2 and Table 31–3 are completely filled in.

TABLE 31-3 Data for pH Experiments

A. Numbers of Dead Animals After 1 Hour

pH	Dead Daphnia	Dead Tubifex
Control		

B. Number of *Daphnia* at Each Level

pH	Level	3 min.	6 min.	9 min.	12 min.	15 min.	1 hour
	Upper						
	Middle						
	Lower						
	Upper						
	Middle						
	Lower						
	Upper						
	Middle						
	Lower						
	Upper						
	Middle						
	Lower						
	Upper						
	Middle						
	Lower						
Control	Upper						
	Middle						
	Lower						

□ Subproblem 3: *Will a combination of high temperatures, low pH, and high insecticide levels have a greater effect together than when administered separately?*

You have studied the susceptibility of two important organisms in the aquatic food chain to two common pollutants. Each pollutant has been found to affect the survival of the

Daphnia and *Tubifex* at different concentrations. A flaw in our investigation, so far, is that we have studied the effects of the variables *independently*.

9. What is wrong with this approach if we do not proceed further? _____

_____

_____

_____

It has been shown that two independent variables, when combined, can sometimes exert an influence *greater than the sum of their individual effects*. This is understandable if you consider the effects of two relatively harmless drugs on the body. If one drug increases heart rate, blood pressure would not necessarily go up substantially because blood vessels would dilate and compensate for the greater volume of blood. If another drug which constricts blood vessels (normally causing minor increases in blood pressure) was introduced at the same time, the combination of mild heart stimulant and mild vessel constrictor might cause a severe increase in blood pressure. The two drugs are acting *synergistically* with one another. Their combined effects are much greater than one would predict, knowing their independent actions.

With this information in mind, prepare the final series of experiments to solve the original problem. Each group of four will plan and share the responsibilities for these experiments.

■ **THE PROBLEM:**   *What are the levels of thermal and chemical pollution (sulfuric acid and insecticide) which are harmful to two representative aquatic food organisms, Daphnia and Tubifex?*

Prepare 100 ml of the LD_{50} solutions of insecticide and sulfuric acid. Prepare 200 ml of the equivalent of deoxygenated water to duplicate conditions produced by thermal pollution and summer high temperature. The maximum temperature of the river was 31°C (88°F). We will use this temperature as our reference point. In order to simulate the virtually anaerobic (oxygen-free) conditions of the river near city X, boil the water to drive out all dissolved oxygen. Use 200 ml of water in a 500 ml Erlenmeyer flask and allow to cool to 31°C.

Designing the Final Experiment

In the space (page 398) describe how your group has decided to proceed. Remember, you are testing to see if the combination of high temperatures and lack of oxygen influences the survival of *Daphnia* and *Tubifex* exposed to insecticide and lowered pH. You already know the individual LD_{50} dilutions in water at room temperature.

10. Should you use the same dilutions in this series of experiments to determine if there

is a synergistic interaction between the variables? _____ Why not? _____

_____

11. What dilutions should you use? _____

_____

Why? _____

_____

_____

    Your controls would include (1) insecticide and lowered pH at room temperature, (2) water at high temperature, and (3) water at room temperature.

Protocol for final experiment

Data from final experiment

REPORT SHEET   Name _____

                                    Instructor's name _____

Summary of Data from All Experiments by Both Pairs

Conclusions

1. What is the solution to the problem?
2. What specific evidence do you have to back up your statement?
3. Have synergistic relationships become evident?
4. Recommendations?

32 Relationship of Observations to Formulation of Generalizations

Symbiotic Relationships of the Frog

In this laboratory period you will find several kinds of animals living inside the frog. It will be tempting to watch these new and interesting organisms and to observe how they move, their morphology, etc. However, this course is designed to teach you to do something more than merely observe. After you have made careful quantitative observations you will be asked to make generalizations about the relationship between the frog and its inhabitants. All observations in science are made with the hope that they will be useful in developing generalizations about our environment.

Facts treated as separate entities and ends in themselves are relatively useless. If you are satisfied with simply observing a parasite in a frog's lung without trying to learn something from it, you are not exhibiting the true scientific spirit of inquiry. There should be, in addition to the visual appreciation of an experience, the desire to understand how the experience fits into the rest of your surroundings.

It is unfortunate that we are not able to allot more time for observation in this course. Those of you who wish to make further observations or perform further investigations with the animals inside the frog will be permitted to do so, and should request this opportunity of your instructor. During this formal laboratory period, however, make sure that you tie together your observations by answering the questions asked on the following pages, right to the end of this series of exercises.

PRELIMINARY INFORMATION

In most cases the organisms living in any habitat have evolved adaptations to their particular environment which prohibit their survival in an ecologically different situation. For example, while nematodes are found in almost every possible *ecological niche* (aspect of the environment), the nematodes living in or near the roots of grass probably could not adapt themselves to living in the roots of cactus in a desert, or in an oak forest, or in a lake, although other nematodes are found in each of these ecological niches.

An interesting ecological niche is found inside living organisms. Animals which live inside other organisms must be adapted to their surroundings in the same manner as any other animal. Virtually every living thing has other organisms living inside its body, or on its outer surface, or both. Almost every internal organ of mammals (and of most other animals) is the habitat of one or another kind of invading organism.

1. List some specific advantages which might accrue to an organism from living in the small intestines of another animal. _____

_____

_____

_____

2. Can you think of any advantage to an invading organism of using the circulatory system rather than the digestive tract as an ecological niche (for example, blood vessels rather than small intestines)?_____

3. List some specific problems which must be overcome by an animal living in the small intestines of another animal (including problems related to survival of the *species* rather than the individual).

_____   _____

_____   _____

_____   _____

4. What additional problems confront an animal living in the blood vessels?_____

_____

Organisms which must live in some sort of an association with other organisms in order to survive are called **symbionts**. Three categories of **symbiosis** (togetherness) are **commensalism, mutualism** and **parasitism**.

In **commensalism** the organism obtains benefit from its relationship with another organism but does not harm or benefit its partner; it is a kind of "silent partner" to its host. An example is the small crab *Pinnixa faba,* which lives inside the shells of living clams and feeds on small organisms sucked into the·clam's mantle cavity, which the clam would not ordinarily eat.

A typical example of **mutualism** is represented by the relationship between various species of green algae and fungi which live together as lichens, often forming a crust-like flat growth on rock and dead trees. The algae, being green, carry on photosynthesis and produce food, some of which is utilized by the fungi. In turn, the fungi, which have a cup-shaped upper region, protect the algae. Mutualism is a relationship in which both partners benefit.

Parasitism is a relationship wherein one organism, the **host**, supplies some sort of benefit, usually food and/or protection, to another organism, the **parasite**, which, as a consequence of its presence in or on the host, causes harm to the host.

Almost every species of organism on the Earth is associated with another in a type of symbiotic relationship. Man, for example, is infected with at least thirty-two different species of nematodes and has numerous protozoan, flatworm, and arthropod parasites. Even the parasites themselves may have **hyperparasites** (parasites of the parasites).

■ PROBLEM I: *Which came first, the parasite or its host?*

5. Make up a hypothesis about how parasites evolved their particular mode of survival. Under your hypothesis discuss the following points as examples to show the reasoning behind your hypothesis.

a. Why is it necessary to assume that parasites were once free-living?

b. Describe several modes of entry by which free-living organisms could gain entry inside animals such as vertebrates. What mechanisms might these invading organisms have in order to overcome the adverse conditions inside the host (for example, acid in the stomach)?

c. What role might insects, such as mosquitoes, have in the infection of a previously unparasitized mammal?

d. Describe an environment which might preadapt free-living animals or plants toward parasitism. For example, what factors in the environment might incline accidentally swallowed organisms to a life of parasitism?

☐ Hypothesis: _____

_____

_____

Discussion

Many scientists have made conjectures about this problem. Some have fed free-living protozoans to animals to determine if the protozoans were preadapted (prepared) to live as parasites. A species of amoeba belonging to the genus *Hartmanella* seems to be evolving toward a parasitic mode of existence. It lives in sewer sludge which has little or no free oxygen (like the mammalian digestive tract). This capacity to be practically **anaerobic** (requiring little or no oxygen for its metabolism) is an example of preadaptation. It is not difficult to imagine *Hartmanella* being swallowed by a bather in polluted water and surviving as a parasite in his digestive tract.

Look for evidence supporting your hypothesis in Problem II of this exercise.

■ **PROBLEM II:**   *Are the symbionts of frogs parasites or commensals?*

The difference between parasitism and commensalism is seldom clear-cut. If there are signs of tissue reactions, damage, or other trauma to the host as a consequence of the presence of the symbiont, it is a parasite. If, however, no such evidence is apparent, one might tentatively suspect that the symbiont is a commensal.

Tissue reactions would include the formation of cysts, which are chambers (usually composed of shiny white or brownish tissue) in which the parasite is encapsulated, or the accumulation of large numbers of white blood cells and lesions (holes) in any organ. Another method of demonstrating parasitism is to find red blood cells or other host tissue in the digestive tract of the suspected parasite.

Examine a frog according to the directions below in order to determine the nature of its symbiotic relationships.

Directions

Obtain a terminally etherized frog on a dissecting pan at the front table (one frog to each pair of students). The frog should be placed ventral side up on the wax. Pick up a fold of loose skin from the midabdomen with your forceps and cut it with the scissors. Insert the blunt end of the scissors into the incision and cut along the midline up to the underside of the chin and down to the anus between the hind legs. Cut at right angles according to Figure 32-1. Carefully cut through the abdominal muscles and make two flaps as shown. Both the abdominal (Figure 32-2) and thoracic cavities should be exposed. Partner 1 should examine the mouth and pharynx, remove the trachea, lungs, esophagus, stomach, and small intestines, and place them in a watch glass filled with 0.7% saline solution (amphibian saline solution).

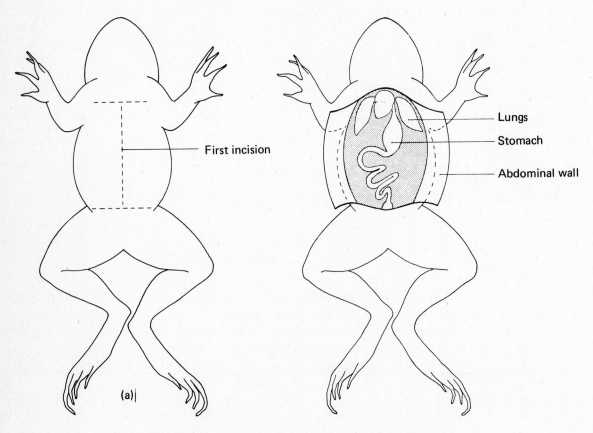

FIGURE 32-1 Ventral view of a frog prepared for dissection.

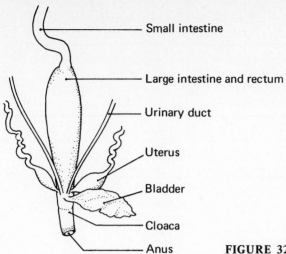

Small intestine

Large intestine and rectum

Urinary duct

Uterus

Bladder

Cloaca

Anus

FIGURE 32-2 Posterior abdominal organs of the frog (lateral view).

Partner 2 should remove the urinary bladder, gall bladder, large intestines, rectum, and cloaca. All organs should be placed in 0.7% saline to prevent drying.

Examine each of the organs you have removed by placing it in a separate watch glass and covering it with 0.7% saline. Tear the organ apart carefully with two dissecting needles, watching for moving objects in the tissues. Do this under a binocular dissecting microscope if possible. When pearly white or brown cysts or organisms are found embedded in the tissues or moving on the bottom of the dish, remove and place them in a separate labeled syracuse dish with 0.7% saline.

The gall bladder and urinary bladder are transparent. They should be removed intact and observed under a binocular dissecting microscope for large worms. If none are found, puncture the bladders and look for protozoans under the highest magnification of the binocular microscope.

If organisms are found, examine the organ from which they came to determine whether or not there are signs of tissue damage. Cysts, usually contributed partly by the host and partly by the parasite, are definite signs of tissue reaction. Other signs are lesions or holes in the tissues which contain one or many of the parasites.

6. Table 32-1 lists the name and location of some of the symbionts found in frogs. Indicate in the appropriate column whether or not the organism is a commensal or a parasite and the evidence for your statement. Confine your answers to those organisms found in your frog.

7. List the symbionts you and your partner found, together with the number of each (estimate when many are present), in Table 32-2. Your instructor will accumulate the information for the class on the chalkboard.

8. Your instructor will poll the class to find out the proportion of infected versus

uninfected frogs. Was any frog completely free of symbionts?_____On the basis of this observation, what can you say about the effectiveness of the mode of transmission of the

symbionts from host to host?_____

_____

TABLE 32-1 Location and Relationships of the Symbionts of the Frog *Rana pipiens*

A. Symbiont	B. Location in Host	C. Hypothetical Relationship with Host (Parasitic, Commensal, Mutualistic)	D. Evidence for C
1. *Pneumoneces* or *Haematolechus* (flatworm)	Lungs; sometimes in mouth or pharynx		
2. *Clinostomum* (flatworm)	In yellowish cysts in mouth or body wall; may escape from cyst during dissection and will then appear as small transparent flatworm with inchworm locomotion		
3. *Rhabdias* (nematode)	Lungs; larvae in small intestines		
4. *Gorgodera* (flatworm)	Urinary bladder		
5. *Cephalogonimus* (flatworm)	Small intestines; has 2 clearly visible suckers: 1 anterior, 1 midventral		
6. *Glypthelmis* (flatworm)	Small intestines; only oral sucker clearly visible		
7. *Diplodiscus* (flatworm)	Large intestine, rectum; one very large posterior sucker		
8. *Opalina* (protozoan)	Large intestine; large, flat, pear-shaped with rows of cilia		
9. *Nyctotherus* and *Balantidium** (protozoans)	In contents of large intestine and rectum; smaller than *Opalina*, ciliated, oval		
10. *Trichomonas** (protozoans)	In contents of large intestine and rectum; very small; move with a wobbly motion; sometimes flagella are visible under high power		

*In order to observe *Nyctotherus*, *Balantidium*, and *Trichomonas* take a drop of the contents of the rectum or large intestine and place it on a slide with 1 drop of 0.7% saline. Place a cover slip on the preparation and observe under low and high-dry lenses of a compound microscope.

TABLE 32–2 Kinds and Quantity of Symbionts in *Rana pipiens*

Symbiont	Number in Your Frog	Average Number per Frog (Whole Class)

9. What might happen to the host if the parasites reproduced at will, gradually building up a very large population?_____

_____

10. Would the same results occur if the population of commensals were so large?_____
Explain. _____

_____

11. Which situation would be more dangerous to the host and why? _____

_____

_____

12. If the relationship between the host and its symbionts were such that the host's ability to survive was interfered with, what influence would this have on the survival of the parasite population?_____

_____

13. If the number of parasites present in the host has an effect on the host's ability to

survive, what inference can you make about control of parasite population levels?_____

_____

_____

14. Would it be an advantage for the host if the parasite population levels were regulated?

_____Why?_____

_____

15. Would it be an advantage for the parasite if its population levels were regulated?_____

Why?_____

_____

16. How can population levels be regulated? Think of as many factors as you can which

affect numbers of organisms._____

_____

_____

_____

_____

17. Have any of the population regulators you mentioned originated in the population

itself, or have they all been a function of the environment?_____

Try to think of ways in which populations are self-regulating._____

_____

_____

_____

18. What population regulators can you think of which are peculiar to the host–parasite
relationship? (For example, if the parasite population builds up, the host is liable to get sick.
A sick host would be weak and unable to get food. This would affect certain parasites—especially those living in the digestive tract and dependent, at least in part, on the host's food—and

reduce their ability to reproduce.)_____

_____

_____

19. The host is the parasite's environment. Compare the interaction between host and parasite in the regulation of parasite population with the effect of the environment on the population of any free-living animal. _____

_____

_____

_____

20. Is there an essential similarity between the mechanisms of regulation of parasite populations and those of free-living organisms? _____

21. Re-examine your list of symbionts of the frog.

    a. List those you are reasonably sure are parasites. _____

_____

_____

    b. List those which might be commensals. _____

_____

_____

    c. Can you be sure of all your decisions? _____ Why or why not?

_____

_____

The problems you have faced today are among the most significant in the field of parasitology. The difficulty you had in deciding whether the symbionts of frogs are commensals or parasites is echoed in many investigations performed by modern parasitologists. For example, the crab *Pinnixa faba* previously referred to lives inside the mantle cavity of the clam *Schizothaerus nuttalii.* It has been considered a commensal, but recently an investigator noticed that it took an occasional bite from the gills of the clam. If this is corroborated, it will be necessary to change its classification from commensal to parasite because it harms its host.

It should be understood, then, that your lack of success in making clear-cut assignments of symbionts as parasites or commensals reflects the difficulty of the problem and not your lack of ability. It should also be understood, at this late point in the course, that scientific method is not a cure-all and does not automatically result in final solutions to all problems. It is limited by the ability of those who would use it and by the nature and difficulty of the problem.

CONCLUSIONS AND GENERALIZATIONS

At the beginning of this exercise you were asked to use observations of the symbionts of frogs to make generalizations about the origin and nature of the parasite–host relationship.

Re-examine the problems you were asked to solve and the observations you made. Make as many generalizations as you can, giving the *evidence* which makes each possible. After each, indicate whether or not you are sure your generalization is valid, and why.

33 Introduction to Analysis

A Field Trip: Trophic Structure of a Rotting-Log Community

All biotic communities consist of a variety of different types of inhabitants tied together into a complex web of interdependencies. Some species are the prey of other species; some are harmless. Yet the fluctuation in the size of a population of any species in a community is sure eventually to affect all the other species, just as changing numbers of policemen, or shopkeepers, or farmers would disrupt a human community.

The thread which binds each species to the others is energy. All ecosystems (except the newly discovered hot water vent community in the deep ocean, which is dependent on sulfur-eating bacteria) depend ultimately on the energy in sunlight for their existence. Each different species has evolved its own particular method for tapping this universal energy flow. The complex combination of life style, place of habitation, type of food, and means of obtaining energy is called the organism's **biological** or **ecological niche**. No two ecological niches in a community can be the same. For if two species of animals, for example, ate exactly the same food under the same circumstances, sooner or later these species would come into conflict—the competition for the same source of energy would result in victor and vanquished—and the vanquished species would become extinct. It is inconceivable that two species would be so alike that neither would have a competitive advantage over the other. So, in the long history of evolution on Earth, niche selection has been of paramount importance. Organisms invaded every possible niche in competition with each other. Gradually some species succumbed to the greater effectiveness of other species, until, of all possible niches, each surviving species found itself with one narrow niche. It was so well adapted to that niche that it outcompeted all the other species. Over thousands or millions of years of evolution the species became more and more finely adapted to its niche until now, if the environment does not change, it is difficult to see how another organism could usurp the niche.

A **community** is a web of niches. Each species is precisely adapted to obtain the energy it needs for survival from the energy budget of the community. A certain amount of sunlight radiates on a community. The green plants convert the sunlight into energy-containing compounds (sugars, etc.). Herbivores eat the plants and use the sugar as a source of energy to keep their life processes going. Carnivores use the energy in the tissues of the herbivores to maintain their life processes. Scavengers and decomposers use the detritus—the leftovers—as a source of energy. They efficiently use the energy that spills over, the extra energy that cannot

411

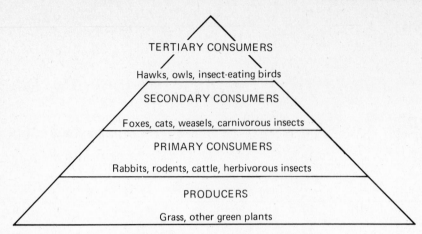

FIGURE 33–1   A trophic pyramid.

be tapped by the other participants in the energy web. For example, after the lions, hyenas, and vultures have finished with a carcass, the flies, beetles, molds, and bacteria use the remnants of organic material until every last drop of accessible energy in the prey animal is consumed.

The energy tied up in a community is usually represented by a **trophic** (energy) **pyramid**. The pyramid shape, a broad base tapering to a point, represents the amount of energy available at each level. The base represents the energy tied up in the green plants or producers, so named because they use the sun's energy to produce energy-rich food. The **primary consumers** or herbivores have the first crack at this vast source of energy, since they eat the plants. The **secondary** and **tertiary consumers** are carnivores; they are more and more remote from the original source of energy, the sunlight, and consequently are able to obtain less and less energy from the system.

Thus grass, converting the sun's energy directly, would have most energy. Rabbits obtain energy from the grass, but some is spilled over and lost, so these primary consumers have less total energy at their level than the producers (grass). Foxes eat rabbits, again losing some energy in the transfer, and hawks might eat an occasional young fox, but certainly, by this time, much less energy is available to the hawk than to the rabbit. The energy tied up in a community of this type is depicted in the trophic pyramid shown in Figure 33–1.

SET THEORY AND ANALYSIS

Today you will be asked to make sense of a jumble of isolated facts. You will examine a community of plants and animals and try to understand the energy relationships among the members of the community. You will be guided through this analysis, but you will need to keep in mind an overview of the community structure as you examine its parts.

Analysis begins with the urge to place isolated facts into a framework and to understand the interactions which occur within that framework. In mathematical terms, the whole is the "set" and the parts comprise "subsets" of the set. But the subsets within the set often share common ground in terms of physical proximity or a functional relationship. Whatever the relationship, the subsets may interact with each other. In biological terms, interactions can occur when one species colonizes an area, preventing other species from living there or crowding them out, or when a tree overgrows an area, placing it into such dense shadow that the growth of the plants beneath its canopy is limited. Or, an interaction can occur when one species eats another.

In the following analysis, the set will be a particular oak forest community; the subsets can be different, depending on the purpose of the analysis. For example, you can represent the community and its constituent populations as

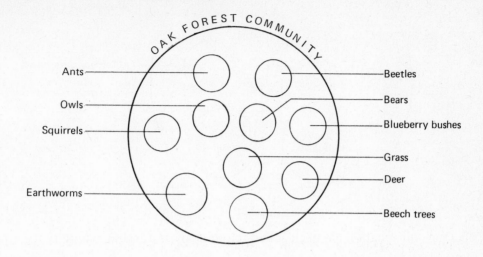

which shows nothing more than the constitutents of the community subsumed under the general heading, oak forest community.

A more informative representation might be

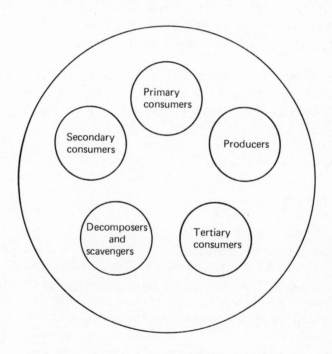

1. What does this set diagram show? _____

_____

Adding still another dimension:

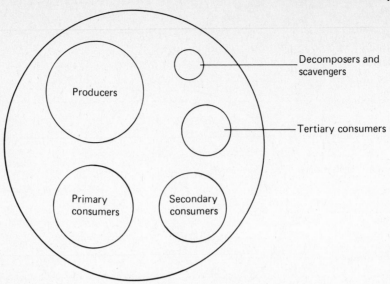

2. What new information has been added to our analysis of the community? _____

_____

Finally, we can diagram the community as follows.

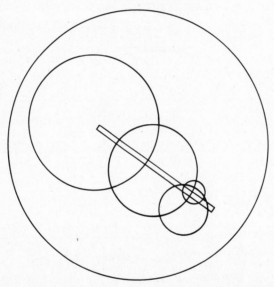

In the above diagram, fill in the terms producer, primary consumer, secondary consumer, tertiary consumer, decomposer. The overlapping areas mean that interactions are occurring.

3. Which of the subsets overlaps only two other subsets? _____

interacts with _____ and _____

4. Which subsets overlap with three other subsets? _____ interacts

with _____ , _____ , and _____ .

Also _____ interacts with , _____ , _____ and

_____. Explain._____

_____

_____

5. Which subset is the long, thin line? _____

Explain what are its interactions and why. _____

_____

_____

6. In the above analysis, relationships between various combinations of sets and subsets were established. Summarize the differences between each of the set diagrams and describe the

general trend. _____

_____

_____

_____

_____

Now you will have an opportunity to analyze a rotting-log community.

THE ROTTING-LOG COMMUNITY

One of the most puzzling biotic communites in terms of trophic (energy) levels is the rotting-log community. Littering the forest floor are blown-down or diseased tree trunks. Each of these becomes the center of a tiny world—a community of organisms which live out their lives in the dark, humid recesses of the rotting tree trunk. It is a veritable jungle, with its eaters and eaten, its herbivores and its carnivores. We cannot transport you to a real jungle or to a coral reef, where spectacular organisms re-enact the perpetual life and death struggle for a share of life-sustaining energy. But today you will have the opportunity to observe the same ecological processes in miniature. Today you will sample a rotting-log community and analyze its trophic composition. In the laboratory there will be a small terrarium tank for each group of four students into which you will put your collected specimens.

Obtain a plastic cup, several squares of paper, and a rubber band. Each group of four students should have one extra cup, to be used as a reservoir. Your instructor will direct you to the rotting log. Use your cup to capture animals by placing it over the animal and then sliding the paper under the cup so that the animal is walking on the surface of the paper. Now invert the cup and the animal will fall to the bottom. Pour your animal into the reservoir and and return to the log for more specimens. The reservoir should be capped with a square of paper fastened with a rubber band.

Each group of four students should attempt to capture all the organisms in 0.3m² (1 ft²) of rotting log.

When in the field, make mental note of the environmental conditions to which the animals were exposed when inside the rotting log.

When you return to the classroom place your animals into the tank and observe and count them.

☐ THE PROBLEM: *To analyze the trophic composition of a rotting-log community.*

☐ Subproblem 1: *To determine the trophic relationships between species.*

☐ Subproblem 2: *To describe the specific ecological niche of each species.*

☐ Subproblem 3: *To predict what would happen to the community if each species, in turn, were to be removed. For example, what would happen in the whole community, over time, if species A were to be completely removed?*

☐ Subproblem 4: *To draw a trophic pyramid using the amount of animal tissue (its mass), to represent the energy tied up in each level.*

In preparing the trophic pyramid you must take into consideration the fact that the amount of energy in a tiny ant is proportionately less than that in a giant beetle. To be able to represent the energy in each, you must convert the beetle into ant-energy units. Thus, as you look at the ant and beetle, you might decide that 100 ants equal 1 beetle in mass. Therefore, the total energy of 100 ants is equal, approximately, to the energy in one beetle. Thus instead of thinking that 100 ants and 2 beetles contain the energy of 102 animals, you would have converted them into 300 ant-energy units. *Remember to list the numbers of organisms found by your group and to convert the mass of these organisms into ant-energy units.*

Use the key on pages 421–424 to identify the types of organisms you have placed in your terrarium tank and fill in Table 33–1. When you have finished your tabulation, place your information on the chart the instructor has drawn on the chalkboard. Then add the data obtained by the class as a whole next to your own numbers on Table 33–1.

Now label the trophic pyramid on page 417 by placing, on the appropriate levels, the names of the organisms in Table 33–1. Notice that the key gives you some clues as to what each type of animal eats.

Using your observations, the data in Table 33–1, and the trophic pyramid, answer the following questions carefully. Do not leave any unanswered.

1. What are the conditions inside the log that compose the environment of the log's in-

habitants? List and/or describe as many environmental factors as you can. _____

_____

_____

_____

	Type of Organism	Number Found	Total Mass in Ant-Energy Units*
1.			
2.			
3.			
4.			
5.			
6.			
7.			
8.			
9.			
10.			
11.			
12.			
13.			
14.	Other		

*If you found 5 isopods, and each represents 50 ant-energy units, the total energy in all 5 isopods would be the equivalent of 250 ant-energy units.

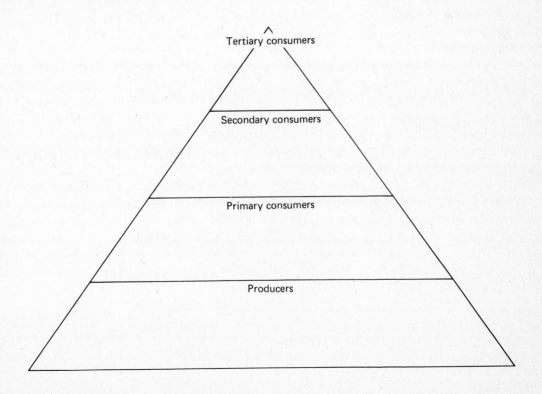

2. It is unlikely that you will have the opportunity to see the animals you have captured in the act of eating. Yet it is of utmost importance to determine what they eat. How else can you decide whether an animal is a primary or secondary consumer? Use the information on the key and the size of each animal's population to determine its trophic level. List the trophic level of each species below. Next to each organism, indicate how the size of its population gives a clue to its trophic level.

| | | *Relationship of Trophic Level* |
| *Species* | *Trophic Level* | *to Size of Population* |

3. Are the inferences you made in question 2 likely to be correct 100% of the time?

_____ Explain. _____

_____

If they are not likely to be completely correct, are they of any use at all? _____

Explain. _____

_____

_____

4. In assigning each species to a trophic level, you will find one startling discrepancy which will immediately strike you. When you tried to fill in the levels of producers, primary consumers, secondary consumers, etc., what did you find conspicuously missing? _____

_____

_____

_____

5. What is the problem facing you when you try to understand the trophic structure of

the rotting-log community, given the fact that green plants are apparently missing? _____

_____

_____

6. Write at least two hypotheses which might explain how the rotting-log community
can be a viable ecological system, despite its apparent peculiarity.

□ Hypothesis 1: _____

_____

□ Hypothesis 2: _____

_____

□ Hypothesis 3: _____

_____

Which hypothesis seems most logical to you? _____

Unfortunately, we have neither the time nor the skills to test your hypotheses to determine which one is correct. We will have to use a technique which leaves much to be desired.
Instead of finding out for ourselves, we will have to use secondary sources in the library. But
how can we look up the answer to our problem? It is unlikely that you will find a book that
discusses the rotting-log community. What will you look up to find out what happened to the
missing trophic level? Describe the topic(s) you will look up below.

7. Based on your reading, was your hypothesis correct? _____ If not, state below the correct answer to the problem of what is peculiar about the rotting-log community energy distribution, and how the community can function despite this peculiarity.

8. Suppose we were to remove all centipedes and carnivorous beetles from the rotting log. What effect would this have on the community? _____

_____

_____

_____

_____

9. Suppose we were to remove all millipedes from the log. What might result in the community? _____

_____

_____

10. What is the "top carnivore" (most important carnivore) in your rotting-log community? _____

11. Describe below how the subsets interact to produce the set "rotting-log community."

Key to the Organisms of the Rotting-Log Community

		Go to Letter

A   Body divided into 3 distinct sections; 3 pairs of legs; 1 pair of antennae

                Phylum Arthropoda, class Insecta   **B**

AA   Body not divided into 3 distinct sections   **H**

B   Head wider than thorax; thorax separated from abdomen by slender "waist"; body brown or red; usually lacks wings; scavenger, eats any plant or animal matter available

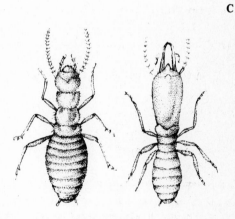

          Order Hymenoptera   Ant

BB   Body otherwise   **C**

C   Head wider than thorax or as wide; body white; usually lacks wings; similar in size to ant; eats wood

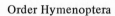

          Order Isoptera   Termite

CC   Body otherwise   **D**

D   Head narrower than thorax or equal in width; 2 pairs of wings, an outer leathery pair and an inner filmy pair; some species carnivorous, some herbivorous, many species eat molds

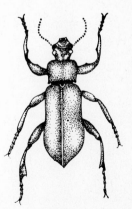

          Order Coleoptera   Beetle

DD   Body otherwise   **E**

Go to Letter

E Rear of body extended into long pincer-like tail;
scavenger, eats plant or animal matter

           Order Dermaptera   Earwig

EE Body otherwise   **F**

 F Head broad, as wide as thorax; may have short wings,
with outer hard pair covering inner filmy pair; last pair
of legs much longer than first 2 pairs; brown or black;
eats plant material

           Order Orthoptera   Cricket

FF Body otherwise   **G**

 G Body appears to have only 2 divisions when viewed from
above, head hidden by dorsal carapace, most of back
covered with chitinous (plastic-like) wings over filmy
wings; eats anything, primarily plant material

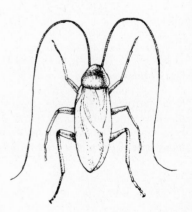

           Order Orthoptera   Cockroach

*Go to
Letter*

GG   With 3 pairs of tiny legs; abdomen long, curved, white, and fleshy; no wings; body curled; eats roots, plant material

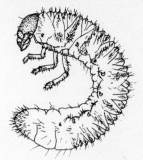

Order Coleoptera   Beetle larva (grub)

H   Possesses coiled shell and soft body; plant eater

Phylum Mollusca, class Gastropoda   Land (pulmonate) snail

HH   Lacks coiled shell   **I**

I   Body soft   **J**

II   Body hard, more than 3 pairs of legs   **K**

J   Looks like a snail without a shell; conspicuous "saddle" on back; 4 tentacles on head; plant eater

Phylum Mollusca, class Gastropoda   Slug

JJ   Elongated, worm-like, conspicuous band (clitellum) about one-third down from head; conspicuously segmented; eats soil, very rotted wood

Phylum Annelida, class Oligocheta   Earthworm

K   Body segmented; more than 4 pairs of legs   **L**

KK   Body in 2 parts, cephalothorax and abdomen; 4 pairs of legs   **M**

Go to
Letter

L   Body flattened, relatively broad; length about 3 times
width; 1 pair of antennae; 7 pairs of legs; body clearly
segmented; eats rotting wood, leaves, etc.

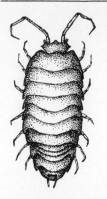

Phylum Arthropoda, class Crustacea, order Isopoda   Isopod (sowbug, pillbug)

LL Body elongate, length more than 5 times width   **M**

M   Body clearly segmented, 1 pair of jointed appendages on
each segment; 1 pair of antennae; carnivorous

Phylum Arthropoda, class Myriapoda, subclass Chilopoda   Centipede

MM Body clearly segmented, 2 pairs of jointed appendages on
each segment; in small specimens [about 2.5 cm (1 in.)]
legs are so tiny they look like fuzz; eats molds, rotting
plants, wood

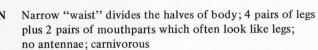

Phylum Arthropoda, class Myriapoda, subclass Diplopoda   Millipede

N   Narrow "waist" divides the halves of body; 4 pairs of legs
plus 2 pairs of mouthparts which often look like legs;
no antennae; carnivorous

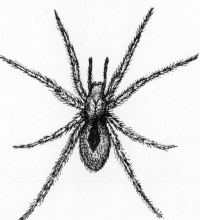

Phylum Arthropoda, class Arachnida, order Araneae   Spider